Artificial Intelligence and the Two Singularities

Chapman & Hall/CRC
Artificial Intelligence and Robotics Series
Series Editor: Roman V. Yampolskiy

The Virtual Mind
Designing the Logic to Approximate Human Thinking
Niklas Hageback

Intelligent Autonomy of UAVs
Advanced Missions and Future Use
Yasmina Bestaoui Sebbane

Artificial Intelligence, Second Edition
With an Introduction to Machine Learning
Richard E. Neapolitan and Xia Jiang

Artificial Intelligence and the Two Singularities
Calum Chace

For more information about this series please visit:
https://www.crcpress.com/Chapman--HallCRC-Artificial-Intelligence-and-Robotics-Series/
book-series/ARTILRO

Artificial Intelligence and the Two Singularities

By
Calum Chace

CRC Press
Taylor & Francis Group
Boca Raton London New York

CRC Press is an imprint of the
Taylor & Francis Group, an **informa** business

A CHAPMAN & HALL BOOK

CRC Press
Taylor & Francis Group
6000 Broken Sound Parkway NW, Suite 300
Boca Raton, FL 33487-2742

© 2018 by Taylor & Francis Group, LLC

CRC Press is an imprint of Taylor & Francis Group, an Informa business
No claim to original U.S. Government works

Printed on acid-free paper

International Standard Book Number-13: 978-0-8153-6853-3 (Paperback)
International Standard Book Number-13: 978-0-8153-6854-0 (Hardback)

Visit the Taylor & Francis Web site at
http://www.taylorandfrancis.com

and the CRC Press Web site at
http://www.crcpress.com

For Julia and Alex

Contents

Acknowledgements

I am grateful to the following people (listed alphabetically), who have all helped the writing of this book. Any mistakes and idiocies are mine, not theirs. I apologise to any I have omitted.

Adam Jolly, Adam Singer, Alex Chace, Alexander Chace, Aubrey de Grey, Ben Goertzel, Ben Goldsmith, Ben Medlock, Brad Feld, Charles Radclyffe, Chris Meyer, Clive Pinder, Dan Goodman, Daniel van Leeuwen, David Fitt, David Shukman, David Wood, Gerald Huff, Henry Whitaker, Hugh Pym, Hugo de Garis, Jaan Tallinn, James Hughes, Janet Standen, Jeff Pinsker, Jim Muttram, Joe Pinsker, John Danaher, Justin Stewart, Keith Bullock, Kenneth Cukier, Leah Eatwell, Malcolm Myers, Mark Mathias, Mike Johnston, Parry Hughes-Morgan, Peter Fenton O'Creevy, Peter Monk, Randal Koene, Russell Buckley, Stuart Armstrong, Stuart Russell, Tess Read, Tim Cross, Tom Aubrey, Tom Hunter, Will Gilpin, William Charlwood, William Graham.

My profoundest thanks go to my partner, Julia Begbie, my staunchest supporter and most constructive critic.

About the Author

Calum Chace is a best-selling author of fiction and non-fiction books and articles focusing on the subject of artificial intelligence. He is a regular speaker on artificial intelligence and related technologies, and runs a blog on the subject at www.pandoras-brain.com. Prior to becoming a full-time writer and speaker, he spent 30 years in business as a marketer, a strategy consultant, and a CEO. A long time ago, he studied philosophy at Oxford University, where he discovered that the science fiction he had been reading since boyhood was simply philosophy in fancy dress.

Introduction

Stock illustration ID: 576208318

The papers these days are full of stories about robots stealing our jobs, and then wiping us out, like the Terminator. This isn't just because our imaginations have been warped by too many science fiction movies; there is a reason for it. The reason is that Artificial Intelligence (AI) has crossed a threshold. It now works, and the machines' capabilities are overtaking ours.

The science of AI was born a little over 60 years ago at a conference in Dartmouth College in the United States. For a long time, it failed to live up to its promise, but earlier this decade, an old approach to AI called neural networks was revitalised and renamed as deep learning. Since then, AI systems have overtaken humans at image recognition (including facial recognition), and are catching up with us at speech recognition and natural language processing.

Almost every day the media reports the launch of a new service, a new product, a new demonstration powered by AI. Some of this is hype, but much of it is real. People are asking themselves, where will it end? The surprising truth is, the AI revolution has hardly begun.

For more than 50 years, computers have obeyed Moore's Law, which observes that the processing power of a computer doubles every 18 months. (Pedants will argue that Moore's Law is dead, and they have a point, but not a very interesting one. Computers are still getting twice as powerful every 18 months or so.) Repeated doubling is exponential growth, and exponential growth is astonishingly powerful. If you take 30 normal steps you will travel around 30 metres. If you could take 30 exponential steps, you would travel to the moon.

To be more accurate, your 29th step would take you to the moon, and your 30th step would bring you all the way back. Each step in an exponential process is equal to the sum of all the previous steps. This means you are effectively always at the beginning. However interesting the past seems, it is nothing compared to what is coming. We humans tend to forget or overlook the impact of exponential growth, and that can be dangerous. When thinking about our future with AI, we must keep it in mind.

This book argues that in the course of this century, the exponential growth in the capability of AI is likely to bring about two 'singularities', a term borrowed from maths and physics, where it means a point at which conditions are so extreme that the normal rules break down.

The first is the economic singularity, when machines render many of us unemployable and we need a new economic system. The second is the technological singularity, when we create a machine with all the cognitive abilities of an adult human, which quickly becomes a superintelligence, and humans become the second smartest species on the planet.

These ideas seem fantastical to many of us at the moment, but the arguments for expecting them to happen are strong, and many of the people best-placed to judge whether they are likely to happen think they will. If and when they do arrive, they will present us with serious challenges, and we need to prepare for them. We should all be reading articles and books about them, thinking about them and discussing their implications. Our future welfare depends on it.

There are many other causes for concern about AI, including privacy, transparency, security, bias, inequality, isolation, oligopoly, killer robots and algocracy. But none of these issues could throw our civilisations into reverse gear, or even destroy us completely. The economic and technological singularities could do that, if we are foolish and unlucky.

The good news is that there are now several organisations dedicated to meeting the challenges posed by the technological singularity. There are too few of them, and they need more funding, but work has at least begun, and we probably have a few decades for them to succeed.

However, there are almost no organisations dedicated to helping us navigate safely through the economic singularity. This is a grave shortcoming: it is becoming urgent that we establish some.

I am confident that we can meet the challenges posed by the two singularities and overcome them. If we do, then AI should turn out to be the best thing ever to happen to humanity, making our future wonderful almost beyond imagination. The economic singularity could lead to radical abundance: societies where poverty is abolished, and we humans are busy doing whatever it is we want to do. The technological singularity could solve all of our current major problems (while probably presenting some interesting new ones), making us almost godlike.

Let's make it so.

I

Artificial Intelligence

CHAPTER 1

An Overnight Sensation, after 60 Years

1.1 DEFINITIONS

1.1.1 Intelligence

Artificial intelligence (AI) can be defined as intelligence demonstrated by a machine or by software. That's easy enough, but what is intelligence?

Like most words used to describe what the brain does, intelligence is hard to pin down, and there are many rival definitions. Most of them contain the notion of the ability to acquire information, and use it to achieve a goal. One of the most popular recent definitions is from German academic Marcus Hutter and Shane Legg, a co-founder of a company called DeepMind that we will hear about later. It states that 'intelligence measures an agent's general ability to achieve goals in a wide range of environments'.[1]

As well as being hard to define, intelligence is also hard to measure. There are many types of information that an intelligent being might want to acquire, and many types of goals it might want to achieve.

An American psychologist called Howard Gardner has distinguished nine types of intelligence: linguistic, logic-mathematical, musical, spatial, bodily, interpersonal, intrapersonal, existential and naturalistic.[2] Just listing them is sufficient for our purposes: we don't need to examine each one. Gardner has been criticised for not providing experimental evidence for these categories, but teachers and thinkers find them very useful.

Certainly, we all know that people vary in the type of intelligence they possess. Some are good at acquiring dry factual knowledge, such as the birth dates of kings and queens, yet hopeless at using their knowledge to achieve goals, like making new friends. Others struggle to learn things from books or lessons, but are quick to understand what other people want, and hence become very popular.

When thinking about intelligence, a host of associated notions crowd in, such as reasoning, memory, understanding, learning and planning.

Intelligence is, of course, the distinguishing feature of humans: it is the characteristic which sets us apart from other animals and makes us more powerful than them. And we are much, much more powerful than them. Genetically, we are almost identical to chimpanzees, and our brains are not much heavier than theirs per kilo of body weight. But the difference in structure between our brains means that there are 7 billion of us and only a few hundred thousand of them.[3] They have no say in their fate, which depends entirely on our decisions and actions: it is probably a blessing for them that they are not aware of that fact.

Our intelligence enables us to communicate, to share information and ideas, and to devise and execute plans of action. It also enables us to develop tools and technology. A single unarmed human would be slaughtered by a mammoth or a lion, but a group of humans working together, or a single human equipped with a rifle, can turn the tables very effectively.

1.1.2 Machine

The second half of our initial definition of AI specified that the intelligence has to be demonstrated by a machine, or by software. The machine in question is usually a computer, although it could be any device created by humans – or indeed by an alien species. Today's computers use processors made of silicon, but in future other materials like graphene may come into play.

('Computer' is an old word which predates the invention of electronic machines by several centuries. Originally, it meant a person who calculates, and in the early twentieth century, companies employed thousands of clerks to spend long and tedious days doing jobs which today's pocket calculators could do in moments.)

Software is a set of instructions that tell electronic signals how to behave inside a machine. Whether intelligence resides in the machine or in the software is analogous to the question of whether it resides in the neurons in your brain or in the electrochemical signals that they transmit and receive. Fortunately, we don't need to answer that question here.

1.1.3 Narrow and General AI

We need to distinguish between two very different types of AI: artificial narrow intelligence (ANI) and artificial general intelligence (AGI[4]), which are also known as weak AI and strong or full AI.

The easiest way to do this is to say that an AGI is an AI system which can carry out any cognitive function that a human can. We have long had computers which can add up much better than any human, and computers which can play chess better than the best human chess grandmaster. But computers are a very long way from being able to beat humans at every intellectual endeavour.

One of the differences between our general intelligence and the machines' narrow intelligence is that we can learn useful lessons from one activity and apply them to another. Learning to play snakes and ladders makes it easier to learn how to play Ludo. A machine that learns snakes and ladders has to forget that in order to learn how to play Ludo. This is called 'catastrophic forgetting', and it is something that AI researchers are trying to overcome.[5]

Another important distinction between narrow AI and AGI involves goal-setting. Narrow AI does what we tell it to. An AGI will have the ability to reflect on its goals and decide whether to adjust them. It will have volition.

1.1.4 Intelligence and Consciousness

Intelligence, the ability to process information and solve problems, is very different to consciousness, the possession of subjective experience. We value intelligence highly, since it is the source of our power, but we value consciousness even more. Most humans are happy to kill and eat animals, which we deem to have a lower level of consciousness than our own.

There is no reason to suppose that humans have attained anywhere near the maximum possible level of intelligence, and it seems highly probable that we will eventually create machines that are more intelligent than us in all respects – assuming we don't blow ourselves up first. We don't yet know whether those machines will be conscious, let alone whether they will be more conscious than us – if that is even a meaningful question. Some people (including me) think that an AGI will probably have self-awareness and be conscious, but this is a hunch rather than a proposition that can be proved yet.

1.1.5 Terminology

Some people dislike the term AI. Pointing out that cars are not called artificial horses, and planes are not called artificial birds, they prefer terms like machine intelligence, or cognitive computing. I sympathise, although for the moment at least, the term artificial intelligence, or AI, is the one understood by the broadest range of people. I will use the terms machine intelligence and AI as synonyms, and sometimes the term 'machines' will encompass AI systems.

1.2 A SHORT HISTORY OF AI RESEARCH

1.2.1 Durable Myths

Stories about artificially intelligent creatures go back at least as far as the ancient Greeks. Hephaestus (Vulcan to the Romans) was the blacksmith of Olympus: as well as creating Pandora, the first woman, he created life-like metal automatons.

More recently, science fiction got started with Mary Shelley's *Frankenstein* in the early nineteenth century, and in the early twentieth century, Karel Capek's play *RUR* (Rossum's Universal Robots) introduced the idea of an uprising in which robots eliminate their human creators.

1.2.2 Alan Turing

The brilliant British mathematician and codebreaker Alan Turing is often described as the father of both computer science and AI. His most famous achievement was breaking the German naval ciphers at the codebreaking centre at Bletchley Park during World War II. He used complicated machines known as bombes, which eliminated enormous numbers of incorrect solutions to the codes so as to arrive at the correct solution. His work is estimated to have shortened the war by 2 years, but incredibly, his reward was to be prosecuted for homosexuality and obliged to accept injections of synthetic oestrogen which rendered him impotent. He died 2 years later and it took 57 years before a British government apologised for this barbaric behaviour.

Before the war, in 1936, Turing had already devised a theoretical device called a Turing machine. It consists of an infinitely long tape divided into squares, each bearing a single symbol. Operating according to the directions of an instruction table, a reader moves the tape back and forth, reading one square – and one symbol – at a time. Together with his PhD tutor, Alonzo Church, he formulated the Church–Turing thesis, which says that a Turing machine can simulate the logic of any computer algorithm.

(The word algorithm comes from the name of a ninth-century Persian mathematician, Al-Khwarizmi.[6] It means a set of rules or instructions for a person or a computer to follow. It is different from a programme, which gives a computer precise, step-by-step instructions how to handle a very specific situation, such as opening a spreadsheet or calculating the sum of a column of figures. An algorithm can be applied to a wide range of data inputs. A machine learning algorithm uses an initial data set to build an internal model which it uses to make predictions; it tests these predictions against additional data and uses the results to refine the model. The way that some game-playing AIs become superhuman in their field is by playing millions of games against versions of themselves and learning from the outcomes.)

Turing is also famous for inventing a test for artificial consciousness called the Turing Test, in which a machine proves that it is conscious by rendering a panel of human judges unable to determine that it is not. Which is essentially the test that we humans apply to each other.

1.2.3 The Birth of Computing

The first design for a Turing machine was made by Charles Babbage, a Victorian academic and inventor, long before Turing's birth. Babbage never finished the construction of his devices, although working machines have recently been built based on his designs. His 'difference engine' (designed in 1822) would carry out basic mathematical functions, and the 'analytical engine' (design never completed) would carry out general-purpose computation. It would accept as inputs the outputs of previous computations recorded on punch cards. Babbage's collaborator, Ada Lovelace, has been described as the world's first computer programmer thanks to some of the algorithms she created for the analytical engine.

The first electronic digital computer was the Colossus, built by codebreakers at Bletchley Park (although not by Turing). But the first general-purpose computer to be completed was ENIAC (electronic numerical integrator and computer), built at the Moore School of Electrical Engineering in Philadelphia, and unveiled in 1946. Like many technological advances, it was funded by the military, and one of its first applications was a feasibility study of the hydrogen bomb. While working on ENIAC's successor, EDVAC (electronic discrete variable automatic computer), the brilliant mathematician and polymath John von Neumann wrote a paper describing an architecture for computers which remains the basis for the great majority of today's machines.

1.2.4 The Dartmouth Conference

The arrival of computers combined with a series of ideas about thinking by Turing and others led to 'the conjecture that every ... feature of intelligence can in principle be so precisely described that a machine can be made to simulate it'. This was the claim of the organisers of a month-long conference at Dartmouth College in New Hampshire in the summer of 1956, which quickly became seen as the foundation event for the science of AI. The organisers included John McCarthy, Marvin Minsky, Claude Shannon and Nathaniel Rochester, all of whom went on to contribute enormously to the field.

In the years following the Dartmouth Conference, some impressive advances were made in AI, and some even more impressive claims were advanced for its potential. Machines were built that could solve high school maths problems, and a programme called Eliza became the world's first chatbot, occasionally fooling users into thinking that it was conscious.

These successes and many others were made possible in part by surprisingly free spending by military research bodies, notably the Defence Advanced Research Projects Agency (DARPA, originally named ARPA), which was established in 1958 by President Eisenhower as part of the shocked reaction in the United States to the Soviet achievement of launching Sputnik, the first satellite to be placed into orbit around the Earth.

The optimism of the nascent AI research community in this period is encapsulated by some startling claims by its leading lights. Herbert Simon said in 1965 that 'machines will be capable, within 20 years, of doing any work a man can do',[7] and Marvin Minksy said 2 years later that 'Within a generation ... the problem of creating "artificial intelligence" will substantially be solved'.[8] These were hugely over-optimistic claims which, with hindsight, look like hubris. But hindsight is a wonderful thing, and it is unfair to criticise harshly the pioneers of AI for underestimating the difficulty of replicating the feats which the human brain is capable of.

1.2.5 Perceptrons and Good Old-Fashioned AI

One of the algorithms which inspired the highest hopes was a simplified model of the neurons in the human brain, invented in 1957 by Frank Rosenblatt. Its instantiation in a machine called a perceptron was the first artificial neural network (ANN). The optimism was quashed in 1969 by a book published by Marvin Minsky and Seymour Papert, two of AI's founding figures. Their book, *Perceptrons*, was interpreted as providing mathematical proof that ANNs had severe limitations. Years later it was shown that these limitations could be overcome by more sophisticated neural networks, but at the time, the critique was devastating.

For the next decade, research was therefore focused on an approach called symbolic AI, in which researchers tried to reduce human thought to the manipulation of symbols, such as language and maths, which could be made comprehensible to computers. This was dubbed good old-fashioned AI, or GOFAI.

1.2.6 AI Winters and Springs

It became apparent that AI was going to take much longer to achieve its goals than originally expected. There were rumblings of discontent among funding bodies from the late 1960s, and they crystallised in a report written in 1973 by mathematician James Lighthill for the British Science

Research Council. A particular problem identified in the Lighthill report is the 'combinatorial problem', whereby a simple problem involving two or three variables becomes vast and possibly intractable when the number of variables is increased. Thus simple AI applications which looked impressive in laboratory settings became useless when applied to real-world situations.

From 1974 until around 1980, it was very hard for AI researchers to obtain funding, and this period of relative inactivity became known as the first AI winter. This bust was followed in the 1980s by another boom, thanks to the advent of expert systems, and the Japanese fifth generation computer initiative. Expert systems limit themselves to solving narrowly defined problems from single domains of expertise (for instance, litigation) using vast data banks. They avoid the messy complications of everyday life, and do not tackle the perennial problem of trying to inculcate common sense.

Japan proclaimed its fifth generation project as the successor to the first generation of computing (vacuum tubes), the second (transistors), the third (integrated circuits) and the fourth (microprocessors). It was an attempt to build a powerful presence for the country in the fast-growing computer industry, and also to abolish the widespread perception at the time that Japanese firms simply made cheap copies of Western-engineered products. The distinguishing feature of the fifth generation project was its adoption of massively parallel processing – the use of large numbers of processors performing coordinated calculations in parallel. Inevitably, Western countries responded by restoring their own funding for major AI projects. Britain launched the £350 million Alvey project in 1983, and the following year, DARPA set up a Strategic Computer Initiative.

Old hands began to fear that another bubble was forming, and they were proved right in the late 1980s when the funding dried up again. The reason was (again) the underestimation of the difficulties of the tasks being addressed, and also the fact that desktop computers and what we now call servers overtook mainframes in speed and power, rendering very expensive legacy machines redundant. The boom and bust phenomenon was familiar to economists, with famous examples being Tulipmania in 1637 (perhaps exaggerated[9]), and the South Sea Bubble in 1720. It has also been a feature of technology introduction since the Industrial Revolution, seen in canals, railways and telecoms, as well as in the dot-com bubble of the late 1990s.

The second AI winter thawed in the early 1990s, and AI research has been increasingly well funded since then. Some people are worried that the present excitement (and concern) about the progress in AI is merely the latest boom phase, characterised by hype and alarmism, and will shortly be followed by another damaging bust, in which thousands of AI researchers will find themselves out of a job, promising projects will be halted and important knowledge and insights lost.

However, there are reasons for AI researchers to be more sanguine this time round. AI has crossed a threshold and gone mainstream for the simple reason that it works. It is powering services which make a huge difference in people's lives, and which enable companies to make a lot of money: fairly small improvements in AI now make millions of dollars for the companies that introduce them. AI is here to stay because it is lucrative. This has happened partly because of big data.

1.2.7 Big Data

Big data was *the* hot topic in business circles in the early 2010s. Businesses and other organisations (especially governments) have massively more data at their disposal than just a few years ago, and it takes considerable effort and ingenuity to figure out what to do with it – if anything.

The term was coined in the mid-1990s by John Mashey at Silicon Graphics, a computer firm, to describe very large and growing data sets which could yield surprising insights into a wide range of phenomena.[10]

Without really trying, we are generating and capturing much more data each year than the year before. It is said that 90% of all the data in existence was created in the last 2 years.[11] The number of cameras, microphones, and sensors of all kinds deployed in the world is growing exponentially, and their quality is improving equally fast. We also leave digital footprints whenever we use social media, smartphones and credit cards.

As well as generating more data, we are quickly expanding our capacity to store and analyse it. Turning big data into information and thence into understanding and insight is the job of algorithms – in other words, of AI.

Big data was helpfully explored in a book of that name published in 2013 by Oxford professor Viktor Mayer-Schönberger and *Economist* journalist Kenneth Cukier. Generally optimistic in tone, it offers a series of case studies of ways in which companies and governments are trawling through oceans of data looking for correlations which enable them to understand and influence the behaviour of their customers and citizens. Airlines can work out the best pricing policy for individual seats on each day before a flight. Hollywood studios can avoid making movies which will lose millions of dollars (and perhaps also avoid making original surprise hits).

Big data has some interesting unexpected side effects. It turns out that having more data often beats having better data, if what you want is to be able to understand, predict and influence the behaviour of large numbers of people. It also turns out that if you find a reliable correlation, then it often doesn't matter if there is a causal link between the two phenomena. We all know of cases where correlation has been mistaken for causation and ineffective or counterproductive policies have been imposed as a result. But if a correlation persists long enough, it may provide decision makers with a useful early warning signal.

Big data also has some negative aspects. Notoriously, government agencies like the National Security Agency (NSA) and Britain's Government Communications Headquarters (GCHQ) collect and store gargantuan amounts of data on us. They claim this is solely to prevent terrorist atrocities, but can they be trusted? They have been less than forthcoming about what information they are gathering, and why. What happens if the data falls into the hands of even less scrupulous organisations?

In fact, it may not be the NSA and GCHQ that we have to worry about. It is reported that they are unable to offer machine learning experts the same salary, lifestyle or moral prestige as Google and the other tech giants. This may be less of a handicap for the security agencies in other countries, for example, the Third Division of China's People's Liberation Army.[12]

But concerns over privacy can themselves lead to perverse outcomes. Cukier argues that the Ebola epidemic of 2014–2015 could have been ameliorated faster – and many lives could have been saved – if phone records had been used to track and analyse the movement of people around West Africa. Fortunately, Ebola did not turn into the pandemic that many feared it would, and Cukier recommends that we urgently review our priorities regarding privacy concerns before that does happen.[13]

Big data was an important phenomenon in its own right, but it achieved particular significance because it helped to make AI an overnight success, after 60 years of trying.

1.2.8 Machine Learning: AI's Big Bang

In the last few years, the field of AI has undergone a quiet revolution. It goes by the name of machine learning, and a subset called deep learning has proved especially effective at tasks which were previously considered hard problems that were unlikely to be solved for many years to come.

Traditionally, statisticians started with a hypothesis and went to look for data to support or refute it. Machine learning, by contrast, starts with data – lots of it – and looks for patterns. Today, as mentioned, with our profusion of smartphones, sensors and tracking devices, we are generating enormous amounts of data.

The turning point came in 2012 when researchers in Toronto, led by Geoff Hinton, won an AI image recognition competition called ImageNet.[14] Hinton is a British researcher now working at Google as well as Toronto University, and the most important figure behind the rise of deep learning as the most powerful of today's AI techniques. His key colleagues at Toronto were Yann LeCun (now at Facebook) and Yoshua Bengio, a professor at the University of Montreal, and now working closely with Microsoft.

The 2012 achievement was the culmination of a process which began back in 1986, when Hinton realised that a process called backpropagation could train a neural network to be effective. Backpropagation retraces the steps in a neural net to identify the errors made in an initial processing of input data. As the name suggests, the error is propagated backwards down the net. It took 26 years until sufficient data and computational power was available to make backpropagation actually work.[15]

1.2.9 Types of Machine Learning

Machine learning can be supervised, unsupervised or reinforcement learning. In supervised machine learning, the computer is given pre-labelled data, and required to work out the rules that connect them. In unsupervised learning, the machine is given no pointers, and has to identify the inputs and the outputs as well as the rules that connect them. In reinforcement learning, the system gets feedback from the environment – for instance by playing a video game.

Machine learning researchers work with a wide range of algorithms. In his 2015 book, *The Master Algorithm*, Pedro Domingos categorises the researchers into five tribes: the symbolists, the evolutionaries, the Bayesians, the analogisers and the connectionists. We have met the *symbolists* already: they are the purveyors of GOFAI.

Evolutionists start with a group of possible solutions to a problem and produce a second 'generation' by introducing small random changes and removing the least effective solutions from the population – analogous to the process of mutation and natural selection described by Darwin. Their champion from the 1960s onward was John Holland, until his death in 2015. Evolutionary computation can be very effective, but its critics argue it is circuitous and slow.

Bayesians employ a theorem developed by a Victorian British mathematician and church minister called Thomas Bayes. A Bayesian network makes hypotheses about uncertain situations and updates its degree of belief in each hypothesis according to a mathematical rule when new evidence is provided. The system generates a flow chart with arrows linking a number of boxes, each of which contains a variable or an event. It assigns probabilities to each of them happening, dependant on what happens with each of the other variables. The variables might be, for instance, missing the last train, spending a night in the open, catching pneumonia and dying. The system tests the accuracy of the linkages and the probabilities by running large sets of data through the model, and ends up (hopefully) with a reliably predictive model.

Analogisers are the least cohesive group, according to Domingos. They rely on the observation that a new phenomenon is likely to follow the same behaviour as the previously observed phenomenon which is most like it. If you are a doctor meeting a new patient, note down her symptoms and then rifle through your files until you find the records of the patient with the most similar symptoms. The more data you have, the more accurate your analogy will be.

1.2.10 Deep Learning

Last, and definitely not least, are the *connectionists*. They are champions of deep learning, which has been the most successful form of machine learning so far.

As we have heard, deep learning is a rebranding of an early approach to AI known as artificial neural nets. Deep learning algorithms use several layers of processing, each taking data from previous layers and passing an output up to the next layer. The nature of the output may vary according to the nature of the input, which is not necessarily binary (just on or off), but can be weighted. The number of layers can vary too, with anything above ten layers seen as very deep learning. In December 2015, a Microsoft team won the ImageNet competition with a system which employed a massive 152 layers.[16]

Yann LeCun describes a typical application of deep learning as follows. 'A pattern recognition system is like a black box with a camera at one end, a green light and a red light on top and a whole bunch of knobs on the front. The learning algorithm tries to adjust the knobs so that when, say, a dog is in front of the camera the red light turns on, and when a car is put in front of the camera the green light turns on. You show a dog to the machine. If the red light is bright, don't do anything. If it's dim, tweak the knobs so that the light gets brighter. If the green light turns on, tweak the knobs so that it gets dimmer. Then show a car, and tweak the knobs so that the red light gets dimmer and the green light gets brighter. If you show many examples of the cars and dogs, and you keep adjusting the knobs just a little bit each time, eventually the machine will get the right answer every time.

Now, imagine a box with 500 million knobs, 1,000 light bulbs and 10 million images to train it with. That's what a typical deep learning system is'.

Deep learning researchers use different algorithms according to the nature of the problem they are looking to solve. Image recognition often calls for convolutional networks, in which the neurons in each layer are only connected to groups of neurons in the next layer. Speech recognition more often uses recurrent networks, in which connections from each layer can loop back to an earlier layer.

In Chapter 2: The State of the Art we will look at the leading players in AI research, and examine its achievements to date.

NOTES

1. http://arxiv.org/pdf/0712.3329v1.pdf.
2. http://skyview.vansd.org/lschmidt/Projects/The%20Nine%20Types%20of%20Intelligence.htm
3. http://www.savethechimps.org/about-us/chimp-facts/.
4. The term AGI has been popularised by AI researcher Ben Goertzel, although he gives credit for its invention to Shane Legg and others: http://wp.goertzel.org/who-coined-the-term-agi/.
5. https://singularityhub.com/2017/03/29/google-chases-general-intelligence-with-new-ai-that-has-a-memory/?utm_content=buffer958e8&utm_medium=social&utm_source=twitter-hub&utm_campaign=buffer
6. http://www.etymonline.com/index.php?term=algorithm.
7. *The Shape of Automation for Men and Management* by Herbert Simon, 1965.
8. *Computation: Finite and Infinite Machines* by Marvin Minsky, 1967.
9. https://www.smithsonianmag.com/history/there-never-was-real-tulip-fever-180964915/.
10. https://setandbma.wordpress.com/2013/02/04/who-coined-the-term-big-data/.
11. http://www.iflscience.com/technology/how-much-data-does-the-world-generate-every-minute/.

12. https://en.wikipedia.org/wiki/People%27s_Liberation_Army#Third_Department.
13. https://www.youtube.com/watch?v=G_HgfrD5tDQ.
14. http://www.wired.com/2016/01/microsoft-neural-net-shows-deep-learning-can-get-way-deeper/.
15. https://www.technologyreview.com/s/608911/is-ai-riding-a-one-trick-pony/.
16. http://www.wired.com/2016/01/microsoft-neural-net-shows-deep-learning-can-get-way-deeper/.

CHAPTER 2

The State of the Art

Artificial Narrow Intelligence

2.1 ARTIFICIAL INTELLIGENCE IS EVERYWHERE

Artificial intelligence (AI) is all around us. People in developed economies interact with AI systems many times every day without being aware of it. If it all suddenly disappeared they would notice, but its omnipresence has become unremarkable, like air.

2.1.1 Smartphones

The most obvious example is your smartphone. It is probably the last inanimate thing you touch before you go to sleep at night and the first thing you touch in the morning. It has more processing power than the computers that NASA used to send Neil Armstrong to the moon in 1969.

In fact, the rate of progress in microprocessors means that observation is now rather old hat: apparently, a modern toaster has more processing power than the Apollo guidance computer.[1] It has been calculated that today's iPhone, if it could have been built in 1957, would have cost one and a half times today's global GDP, would have filled a 100-storey building three kilometres long and wide and would have used 30 times the world's current power generating capacity![2]

Your smartphone uses AI algorithms to offer predictive text and for speech recognition, and these features improve month by month. Many of the apps we download to our phones also employ AI to make themselves useful to us, and of course our phones and apps call on AI systems running elsewhere ('in the cloud') to provide us with some of their most powerful features. The AI in our phones becomes more powerful with each generation of phone as their processing power increases, the bandwidth of the phone networks improve, cloud storage becomes better and cheaper and we become more relaxed about sharing enough of our personal data for the AIs to 'understand' us better.

Many people in the developed economies make several Internet searches a day: at the time of writing, Google carries out 40,000 searches every second.[3] Many of them are performed with the help of AI.

2.1.2 Logistics, Recommendations

When you visit a supermarket – or any other kind of store for that matter – the fact that the products you want are on the shelf is significantly due to AI. The supermarkets and their suppliers are continually ingesting huge data feeds and using algorithms to analyse them and predict what we will collectively want to buy, when and where. The retail supply chain is enormously more efficient than it was even a decade ago thanks to these algorithms.

Other consumer-facing companies like Amazon and Netflix wear their AI on their sleeve, tempting us with products and movies based on their algorithms' analysis of what we have chosen in the past. This is the same principle as direct marketing, which has been around for decades, of course. Nowadays the data available and the tools for analysing it are much better, so people living in skyscraper apartments no longer receive junk email about lawnmowers.

The financial markets make extensive use of AI. High-frequency trading, where computers trade with each other at speeds no human can even follow – never mind participate in – took off in the early twenty-first century, although it has reportedly fallen back from around two-thirds of all US equity trades at the start of the 2008 credit crunch to around 50% in 2012.[4] There is still confusion about the impact of this on the financial markets. The 'flash crash' of 2010, in which the Dow Jones lost almost 10% of its value in a few minutes was initially blamed on high-frequency trading, but later reports claimed that the AIs had actually mitigated the fall. The crash prompted the New York Stock Exchange to introduce 'circuit breakers' which suspend trading of a stock whose price moves suspiciously quickly. The financial Armageddon which some pundits forecast has not arrived, and although there will undoubtedly be further shocks to the system, most market participants expect that new AI tools will continue to be developed for and absorbed by what has always been one of the most dynamic and aggressive sectors of the economy.

Hospitals use AI to allocate beds and other resources. Factories use robots – controlled by AI – to automate production and remove people from the most dangerous jobs. Telecommunications companies, power generators and other utilities use it to manage the load on their resources.

Even though you may not see it, AI is everywhere you look. The world's biggest and most profitable companies are increasingly basing their entire existence on it.

2.2 THE TECH GIANTS

The science of AI is advancing rapidly, with significant steps announced almost every month. Enormous resources are being devoted to achieving these advances. Much cutting-edge work in AI goes on in universities, but much of it also happens inside the tech giants on the US West Coast. Four of them – Intel, Microsoft, Google and Amazon – are among the world's top ten research and development (R&D) spenders, with a combined budget in 2015 of $42 billion.[5] This equals the entire R&D spend of the UK, both public and private.[6] IBM, Apple and Facebook are not far behind, and are increasing their R&D spend sharply.

2.2.1 Google and DeepMind

Google is an AI company. Founded by Larry Page and Sergey Brin in 1998, it makes most of its money (and it makes a phenomenal amount of money!) from intelligent algorithms which match adverts with readers and viewers, and it is busily looking for more and more new ways to exploit its world-leading expertise in AI in as many industries as it can manage. The huge collection of servers which comprise the distributed computing platform for the AI which drives the company's numerous services is often called the Google Brain.

Sometimes Google enters a new industry using home-grown talent, as with its famous self-driving cars, and with Calico, which is looking to apply big data to healthcare. Other times it acquires companies with the expertise not already found inside Google, or 'acqui-hires' their key talent. Its rate of acquisition reached one company a week in 2010, and by the end of 2014 it had acquired 170 of them. Significant industries where Google has engaged by acquiring include smartphones (Android, Motorola), voice over IP telephony (GrandCentral, Phonetic Arts), intelligent house management (Nest Labs, Dropcam and Revolv), robotics (eight robot manufacturers acquired in 2013 alone), publishing (reCAPTCHA and eBook Technologies), banking (TxVia), music (Songza) and drones (Titan Aerospace).

Google's ambitions are startling, and the hurdle for potential acquisition targets is high: they must pass a 'toothbrush test', meaning that their services must be potentially useful to most people once or twice every day.

Google also buys companies (and acqui-hires people) which have expertise in AI that is not already applied to a particular industry. Most famously, it paid more than $500 million in January 2014 for DeepMind, a 2-year-old company employing just 75 people which developed AIs that played video games better than people. At the time of the deal, DeepMind not only had no profits; it had no revenues. As we will see later, it went on to make astonishing contributions to the field.

Later in the year, Google paid another eight-figure sum to hire the seven academics who had established Dark Blue Labs and Vision Factory, two more AI start-ups based in the UK. Before that, in March 2013, it had hired Geoff Hinton, one of the pioneers of machine learning (ML), based in Toronto.

In December 2012, Google hired the controversial futurist Ray Kurzweil as a director of engineering. Kurzweil, of whom more later, believes that artificial general intelligence (AGI or strong AI) will arrive in 2029, and that the outcome will be very positive.

In 2015, Google reorganised, establishing a holding company called Alphabet. Google, the search engine ad business, is by far the largest component by revenue, but the founders spend most of their time on the newer businesses. Most people still refer to the whole company as Google, a convention that I follow here.

2.2.2 Evil Monopoly or Impatient Futurists?

Google's unofficial slogan was 'Don't be evil', which in 2015 was changed to 'Do the right thing'.

Many people are cynical about this, believing that Google is simply trying to disguise its corporate greed in philanthropic clothing. They accuse it of infringing its users' privacy, dodging taxes and being complicit in state censorship in China and elsewhere.

I have no privileged information, but my sense is that Google's founders have a genuine enthusiasm for the future: they think the future will be a better place for humans than the present, and they are impatient for it to arrive.

Philanthropy is a long-established driver for America's business elite – more so than in most other countries. Even the ruthless 'robber barons' of the early twentieth century who made huge fortunes from railroads, commodities and power used their money to establish major benevolent institutions. One of the most successful of those barons was the Scottish-American steel magnate Andrew Carnegie. He pursued wealth aggressively in his early years and spent the last third of his life giving it all away, endowing some 3,000 municipal libraries, and provided funding for several universities and numerous other organisations before he died. His most famous motto was that 'the man who dies rich dies disgraced'.[7]

Bill Gates, Mark Zuckerberg and others seem to be following the same path today, but with a twist. The early industrialists saw themselves as creating wealth, and laying down the infrastructure for modern economies.

The pioneers of the information revolution seem to want to accelerate the progress of technological innovation and use it to transform what it means to be human.

In Google's case, part of this is the ambition to build an artificial brain – an AGI. In May 2002, Larry Page said: 'Google will fulfil its mission only when its search engine is AI-complete. You guys know what that means? That's artificial intelligence'.[8]

2.2.3 America's Other AI Giants

Facebook, Apple, Microsoft, Amazon and IBM are working hard to keep up with Google in the race to develop better and better AI.

Facebook lost out to Google in a competition to buy DeepMind, but in December 2013, it had hired Yann LeCun, a colleague of Geoff Hinton's at the forefront of deep learning. It then announced the establishment of Facebook AI Research (FAIR), which LeCun would run. FAIR has quickly become one of the world's most respected AI labs, responsible for major advances in image recognition systems in particular. According to Joaquin Candela, director of engineering for the company's applied machine learning group (AML), 'Facebook today cannot exist without AI. Every time you use Facebook or Instagram or Messenger, you may not realize it, but your experiences are being powered by AI'.[9]

For many people, the embodiment of AI today is Siri, Apple's digital personal assistant that first appeared preloaded in the iPhone 4S in October 2011. The name is short for Sigrid, a Scandinavian name meaning both 'victory' and 'beauty'. Apple obtained the software in April 2011 by buying Siri, the company which created it, an offshoot of a Defence Advanced Research Projects Agency (DARPA)-sponsored project. Google responded a year later by launching Google Voice Search, later renamed Google Now. Microsoft has Cortana, and Amazon has Alexa, of which more later.

(In May 2016, the original creators of Siri unveiled Viv (from the Latin for 'life').[10] Their company, Six Five Labs, was acquired five months later by the giant Korean company, Samsung, for use in its phones.)

Apple's efforts to keep up with the leaders in AI are widely thought to have been hampered by the firm's notorious culture of secrecy. ML researchers have usually spent some time in academia, and they want to publish and share their findings. Until recently, Apple forbade this, but it changed the rules in December 2016, and in July 2017, it even launched an ML blog.[11] At the time of writing (mid-2017), Apple's digital assistant technology is regarded as decidedly inferior to its rivals, but it is the world's largest company by market capitalisation, and it has enormous reserves to invest in catching up.

Microsoft traditionally adopted a fast-follower strategy rather than being a cutting-edge innovator. In the twenty-first century it seemed to fall behind the newer tech giants, and was sometimes described as losing relevance. When Satya Nadella became CEO in February 2014, he made strenuous efforts to change this perception, and in July that year, Microsoft unveiled Adam, its answer to the Google Brain, running on the company's Azure cloud computing resource.

During 2016 and 2017, Microsoft executed a public pivot towards AI. Announcing a new business unit, the Microsoft AI and Research Group, comprising 5,000 staff in September 2016, Nadella said 'we are infusing AI into everything we deliver'.[12] Just one year later, the company announced that the group had expanded to 8,000.[13] In July 2017, Microsoft announced the formation of a 100-strong lab of AI researchers called Microsoft Research AI working on general-purpose AI, and clearly aiming to challenge the lead opened up by Google and DeepMind.[14]

In January 2013, Amazon bought Ivona, a Polish provider of voice recognition and text-to-speech technology. The technology was deployed in various Amazon products including Kindles, and its unsuccessful launch into mobile phones. In November 2014, the company announced Echo, a smart speaker with a digital assistant called Alexa, which helps users select music and podcasts and make to-do lists, and provides weather and other real-time information. Although Amazon does not break out its sales data, Alexa was a very successful product, with an estimated five million units sold in 2016.[15] Observers were surprised at how sophisticated the AI was, and when Apple launched a competitor called HomePad in June 2017, it competed on the basis of sound quality rather than the capabilities of its digital assistant.

Amazon's founder Jeff Bezos noted in his 2016 letter to shareholders that 'much of what we do with machine learning happens beneath the surface. Machine learning drives our algorithms for demand forecasting, product search ranking, product and deals recommendations, merchandising placements, fraud detection, translations and much more ... quietly but meaningfully improving core operations'.

Named after the company's first CEO, IBM's Watson is a question answering system that ingests questions phrased in natural language, and applies knowledge representation and automated reasoning to return answers, also in natural language. It was developed from 2005 to 2010 in order to win the Jeopardy quiz game, which it did the following January (see subsequent section), to great acclaim. Watson's architecture comprises a collection of different systems and capabilities, including some employing ML techniques.[16]

Since then, IBM has applied the Watson brand to a growing list of applications, notably in the medical sector. In April 2016, the company said that its cognitive computing business accounted for over a third of its $81 billion-annual revenues, and was the main focus for the company's growth,[17] having invested $15 billion in the endeavour since 2010. Watson has announced a host of partnerships with large enterprises, but IBM has had to defend itself against criticisms of being heavily reliant on consulting backup,[18] and also of being more about branding than genuinely cutting-edge AI capability.[19]

2.2.4 China's AI Giants

China has three AI giants, collectively initialised as BAT. Baidu is often described as China's Google, and Alibaba as its Amazon. The third, Tencent, is a phone and Internet giant and is often compared to Facebook.

Baidu is the leading search engine in China, with 56% of the market in 2014.[20] It was founded in 2000, two years after Google, by Robin Li and Eric Wu. In May 2014, it drew attention to its ambitions for AI by hiring Andrew Ng, one of the founders of Google Brain, to head up its new AI lab in Silicon Valley, with a claimed budget of $300 million over five years. Ng had been a leading figure in the development of Google Brain and then went on to help found Coursera, Stanford's online education venture. In the following two years, Baidu spent around $1.5 billion on its AI capabilities, building up a team of 1,200 researchers. Ng left Baidu again in early 2017.

Baidu has been investing heavily in self-driving vehicle technology since at least 2015. It has a partnership with Nvidia (a chip manufacturer that we will hear more of later), and an alliance called Apollo with 50 other technology companies, including Microsoft. Baidu has a declared goal of introducing fully autonomous vehicles in 2020.[21] It will be a key player in the Chinese government's drive to become the leading force in global AI by 2030.[22]

Founded by Jack Ma in 1999, Alibaba overtook Walmart in 2016 to become the world's largest retailer. Its online sales exceed those of Walmart, Amazon and eBay combined. In 2017, it overtook Tencent to become Asia's largest company, with a market cap of $390 billion.[23] Its cloud computing service Aliyun (similar to Amazon Web Services) launched an AI platform called DT PAI in 2015, and in 2017, the company launched a speaker with a digital assistant called Tmall Genie, which drew comparisons with the Amazon Echo. In October 2017, it announced plans to invest more than $15 billion over the following 3 years to research emerging technologies including AI and quantum computing.[24]

Tencent was founded in 1988, and its most important services are social media and games. Its messaging app WeChat has a similar number of users as Facebook Messenger and WhatsApp, but a richer set of functionalities, including a very popular payments service. Half its users spend more than 90 minutes a day on the app. At the time of writing, it had not succeeded in penetrating any major non-Chinese markets. Tencent established an AI lab in its home city of Shenzhen in 2016, with 50 AI researchers and 200 engineers. In 2017, it announced another AI lab in Seattle, headed by senior AI researcher Yu Dong, poached from Microsoft.

AI is a top priority for China's government as well as its companies. AlphaGo's victories in 2016 and 2016 made a huge impact in Korea and China, where Go originated, and where it is still hugely popular. It has been described as China's 'Sputnik moment', after the Soviet satellite launch in 1957 which ignited the space race, and spurred the United States to launch the Apollo programme to land a man on the moon, as well as DARPA's spending on technologies including AI.

China had already become a major force in AI development: a White House report stated in 2016 that the United States had been overtaken by China in the number of academic papers being published on AI each year.[25] In July 2017, China's State Council issued the 'Next Generation Artificial Intelligence Development Plan', which called for China's AI research to be keeping pace with the most advanced labs in the world by 2020, to be making major breakthroughs by 2025, and to be the world's premier location for AI research by 2030.[26]

Shanghai is the focus for much of this activity, with a million square metres of the Huangpu River in its Xuhui district earmarked for AI companies. The 200-metre-tall AI Tower has already been completed on the Longyao Road there, to house multinational AI companies and host exhibitions.[27]

2.2.5 Europe's AI Giants

There are none. DeepMind, based in London, is probably the biggest collection of deep learning researchers in the world, and has been responsible for many impressive achievements by AI in the last few years. But it is owned by Google.

As we will see in Chapter 14: The Challenges, this could be a problem.

Other economic powerhouses are also lagging behind. India set up a task force in August 2017 to draw up policies to promote AI development, and its use in government and elsewhere.[28]

2.2.6 Teenage Sex

Google, Facebook and the other tech giants pioneered the use of ML, and for a while they were pretty much the only organisations with the expertise, the computing resources and the data to implement it. The joke was that ML was like teenage sex – everyone talked about it, but pretty much no one did it. That is changing.

In September 2015, Google announced an important change in strategy. Having built a very lucrative online advertising business based on proprietary algorithms and hardware which produced better search results than anyone else, it was open sourcing its current best AI software – a library of software for ML called TensorFlow.[29] The software was initially licensed for single machines only, so even very well-resourced organisations weren't able to replicate the functionality that Google enjoys. In April 2016, that restriction was lifted.[30]

In October 2015, Facebook announced that it would follow suit by open sourcing the designs for Big Sur, the server which runs the company's latest AI algorithms.[31] Then in May 2016, Google open sourced a natural language processing programme playfully called Parsey McParseFace, and SyntaxNet, an associated software toolkit. Google claimed that in the kinds of sentences it could be used with, Parsey's accuracy was 94%, almost as good as the 95% score achieved by human linguists.[32]

Open sourcing confers a number of advantages for the tech giants. One is to create a level of goodwill among the AI community. Another is the fact that researchers in academia and elsewhere will learn the systems, and be able and inclined to work closely with Google and Facebook – and indeed be hired by them. Also, having more smart people working with their systems means there are more suggestions for improvement and debugging.

2.2.7 Machine Learning Extends beyond the Tech Giants

2015 was the year of the 'great robot freak-out'. Statements in the previous year by the 'three wise men' – Stephen Hawking, Elon Musk and Bill Gates – alerted journalists to the possibility that strong AI might be developed in the foreseeable future, and be followed by superintelligence. The media responded by publishing lots of pictures of the Terminator, which was attention-grabbing, if misleading. After a pause for breath, business leaders got the idea that AI was now a technology that worked, and they began to wonder what it could do for their businesses. Indeed, they began to wonder what it would do *to* their industries, and whether that represented a threat.

Coca-Cola is typical of many of the world's largest companies in that around 2010 it cottoned on to the potential value of analysing the big data it generates about its customers, and in the last couple of years it has been experimenting with applications of ML. It is currently reported to be working on how to install a virtual assistant in its vending machines.[33]

The new awareness of AI and the new availability of sophisticated ML tools have combined to create a thirst for information about what ML is and how it can be used in businesses and other organisations today. A minor industry of conferences (like AI-Europe[34]) and consultancies (like Satalia,[35] Rainbird[36] and Crowdflower[37]) sprang up or morphed their offer in order to satisfy this demand. At the time of writing (autumn 2017), the applications of most interest to business audiences are chatbots and robotic process automation.

Most of the tech giants listed previously sell their products and services to consumers rather than businesses. Much of Microsoft's revenue comes from enterprises, but mostly from identical software packages. IBM is different: it generates much of its revenue from bespoke services to corporates, including consultancy – it acquired PwC's consultancy arm in 2002. As discussed previously, IBM is heavily promoting AI-powered services under the Watson brand.

PwC and the other audit-based consultancies like Ernst & Young and KPMG, and other nonaudit-based consultancies like Accenture are also investing heavily in capabilities to help their clients deploy AI in their organisations.

The intense interest in AI has also produced a thriving ecosystem of start-ups. In mid-2017, research firm Venture Scanner was tracking 1,888 AI companies in 13 categories across 70 countries, with a combined total of $19 billion in funding.[38] But it is still early days for the sector. By one count there were over 300 venture capital deals in AI-based companies during 2015, but 80% of them were for less than $5 million, and 75% of them were in the United States.[39]

2.3 HOW GOOD IS AI TODAY?

2.3.1 Games and Quizzes: Chess

The game of chess used to be thought of as one of the most challenging intellectual pursuits a person could undertake. (Being rubbish at it, I still do.) It used to be thought that it would take centuries for machines to become really good at it. That was a long time ago, of course, and we are much wiser now, because as long ago as 1997, IBM's Deep Blue beat Garry Kasparov, the world's best player, in a controversial but conclusive match. Nowadays, humans have no chance against a chess-playing programme on a smartphone.

2.3.2 Jeopardy

IBM's next bravura AI performance came in 2011, when a system called Watson beat the best human players of the TV quiz game 'Jeopardy', in which contestants are given an answer and have to deduce the question. Watson used 'more than 100 different techniques … to analyze natural language, identify sources, find and generate hypotheses, find and score evidence and merge and rank hypotheses'. It had access to 200 million pages of information, including the full text of Wikipedia, but it was not online during the contest. The difficulty of the challenge is illustrated by the answer, 'A long, tiresome speech delivered by a frothy pie topping' to which the target question (which Watson got right) was 'What is a meringue harangue?' After the game, the losing human contestant Ken Jennings famously quipped, 'I for one welcome our new robot overlords'.[40]

In Chapter 1: An Overnight Sensation, after 60 Years, we noted that intelligence is not a single, unitary skill or process. The fact that Watson is an amalgam – some would say a kludge – of numerous different techniques does not in itself mark it out as different and perpetually inferior to human intelligence. It is nowhere near an AGI which is human level or beyond in all respects. It is not conscious. It does not even know that it won the Jeopardy match. But it may prove to be an early step in the direction of AGI.

2.3.3 Go

In January 2016, an AI system called AlphaGo developed by Google's DeepMind beat Fan Hui, the European champion of Go, a board game. This was hailed as a major step forward: the game of chess has more possible moves (35^{80}) than there are atoms in the visible universe, but Go has even more – 250^{150}.[41] AlphaGo uses a combination of techniques: it started by ingesting a database of 30 million moves in human-played games, and then used reinforcement learning to improve while playing many games against iterations of itself. Finally, it used a Monte Carlo search technique to select the best move in real games.

A match against one of the world's top-ranked (nine-dan) players, the Korean professional Lee Se-Dol, followed in March 2016. Se-Dol was confident, believing it would take a few more years before a computer could beat him. Most computer scientists agreed with him. He was genuinely shocked to lose the series four games to one, and observers were impressed by AlphaGo's sometimes unorthodox style of play. AlphaGo's achievement was another landmark in computer science, and perhaps equally a landmark in human understanding that something important is happening, especially in the Far East, where the game of Go is far more popular than it is in the West: it was reported that 100 million people watched the games. As mentioned previously, this was a Sputnik moment for China.

AlphaGo's final contest against humans took place a year later against the Chinese world number one player, Ke Jie. AlphaGo won 3–0, and afterwards, DeepMind declared it was going into retirement, with its team moving onto other projects.

2.3.4 E-Sports

In August 2017, a system developed by Elon Musk's AI start-up Open AI beat a human professional player of an online battle game called Dota 2. Although the contest was not staged in the most complex version of the game, Musk declared that the game was 'vastly more complex than traditional board games like chess and Go'.[42]

2.3.5 Self-Driving Vehicles

Another landmark demonstration of the power of AI began inauspiciously in 2004. DARPA offered a prize of $one million to any group which could build a car capable of driving itself around a 150-mile course in the Mojave Desert in California. The best contestant was a converted Humvee named Sandstorm which got stuck on a rock after only seven miles.[43] Thirteen years later, Google's self-driving cars have driven well over three million miles without being unambiguously responsible for a single accident. It is true that they have been rear-ended by human drivers a few times, but this is because they obey traffic regulations scrupulously, and we humans are not used to other drivers doing that. A Google car drove into a bus on Valentine's Day 2016, but the facts of that incident remain somewhat ambiguous.[44]

After initial scepticism, the world's automotive manufacturers are now scrambling to master the technologies involved in producing self-driving cars. We will explore this in more detail in Section III of this book.

2.3.6 Search

We are strangely nostalgic about the future, and we are often disappointed that the present is not more like the future that was foretold when we were younger. The year 2015 was the 30th anniversary of the 1985 movie *Back to the Future,* and it was also the year to which the hero travels at the end of the story. PayPal founder Peter Thiel summed up much of the commentary when he lamented that 'we were promised flying cars and instead what we got was 140 characters' (i.e. Twitter).

We didn't get hoverboards, but we did get something even more significant. As recently as the late-twentieth century, knowledge workers could spend hours each day looking for information. Today, less than 20 years after Google was incorporated in 1998, we have something close to omniscience. At the press of a button or two, you can access pretty much any knowledge that humans have ever recorded. To our great-grandparents, this would surely have been more astonishing than flying cars.

(Some people are so impressed by Google Search that they have established a Church of Google, and offer nine proofs that Google is God, including its omnipresence, near-omniscience, potential immortality and responses to prayer.[45] Admittedly, at the time of writing, there are only

427 registered devotees, or 'readers', at their meeting place, a page on the Internet-community site Reddit.[46])

The flying cars may not be so far away. At the time of writing, six serious projects are under way to launch them within a few years. One of the most plausible is Uber's, which plans to fly people between cities using vertical take-off vehicles powered by multiple electric motors. They will take off and land on the roofs of tall buildings, and fly in designated corridors at designated heights.[47]

In the early days, Google Search was achieved by indexing large amounts of the Web with software agents called crawlers, or spiders. The pages were indexed by an algorithm called PageRank, which scored each web page according to how many other web pages linked to it. This algorithm, while ingenious, was not itself an example of AI. Over time, Google Search has become unquestionably AI-powered.

In August 2013, Google executed a major update of its search function by introducing Hummingbird, which enables the service to respond appropriately to questions phrased in natural language, such as, 'what's the quickest route to Australia?'[48] It combines AI techniques of natural language processing with colossal information resources (including Google's own Knowledge Graph, and of course Wikipedia) to analyse the context of the search query and make the response more relevant. PageRank wasn't dropped, but instead became just one of the 200 or so techniques that are now deployed to provide answers. Like IBM Watson, this is an example of how AI systems are often agglomerations of numerous approaches.

In October 2015, Google confirmed that it had added a new technique called RankBrain to its search offering. RankBrain is a ML technique, and it was already the third-most important component of the overall search service.[49] Initially, it was applied to the 15% of searches which comprise words or phrases that have not been encountered before, and converts the language into mathematical entities called vectors, which computers can analyse directly. Microsoft also uses ML techniques in its Bing search engine.

In February 2016, Google announced an important change of leadership in its search division: Amit Singhal was replaced by John Giannandrea.[50] Singhal had overseen the introduction of RankBrain, but was seen as having a bias against applying ML techniques to search because it is often impossible to know how the machine has reached its conclusions. Giannandrea has no such reservations: in his previous role, he oversaw Google's entire AI research activity, including deep learning. This succession is perhaps an allegory of the way that AI is taking over the Internet, on the way to taking over everything else.

One of the benefits Google hopes to obtain by increasing its use of AI in search is extra ammunition in its competition with Amazon. Google's competitors in search are not Microsoft's Bing, and certainly not Yahoo. Thirty-nine per cent of purchases made online begin at Amazon, compared with 11% at Google.[51] Improving that ratio is a key aim for the search giant. We have seen before, with the relative decline of seemingly invincible goliaths like IBM and Microsoft, how fierce and fast-moving the competition is within the technology industry. This is one of the dynamics which is pushing AI forward so fast and so unstoppably.

2.3.7 Image and Speech Recognition

Deep learning has accelerated progress at tasks like image recognition, facial recognition, natural speech recognition and machine translation faster than anyone expected. In 2012, Google announced that an assembly of 16,000 processors looking at 10 million YouTube videos had identified – without being prompted – a particular class of objects. We call them cats.[52] Two years later, Microsoft researchers announced that their system – called Adam – could distinguish between the two breeds of corgi dogs.[53] (Queen Elizabeth is famously fond of corgis, so Adam's skill would be invaluable in certain British social circles.)

In February 2015, Microsoft announced that its AI systems could identify an image better than humans according to the tests laid down by ImageNet, the world's top image recognition competition.[54] A few days later, Google announced it had done even better.[55] Not to be left out, Facebook posted an impressive demonstration video in November 2015.[56]

In January 2016, Baidu (often described as China's Google) showed off a system called DuLight which uses a camera to capture an image of something in front of you, sends the image to an app on your smartphone, which identifies the object and announces what it is. One application of this is to help blind people know what they are 'looking' at.[57] You can download a similar app called Aipoly for free at iTunes.[58]

We humans are very good at recognising each other's faces. Throughout history, it has been vitally important to distinguish between members of your own group who will help you, and members of rival groups who may try to kill you. A Facebook AI system called DeepFace reached human-level ability to recognise human faces in March 2014, scoring 97% in a test based on a database of celebrity photos called Labelled Faces in the Wild

(LFW).[59] The following year, it announced the ability to recognise faces even when they are not looking towards the camera, with 83% reliability. Google now offers the same functionality to users of Google Plus.[60]

Speech recognition systems that exceed human performance will be available in your smartphone soon.[61] In August 2017, Microsoft announced that it had got the word-error rate of its speech transcription system down to 5.1%, the same level as humans. The system's performance is degraded when dealing with noisy environments, and accents it has not encountered before, but progress is being made on these points.[62]

Speech recognition will be highly significant when combined with translation services. Microsoft-owned Skype introduced real-time machine translation in March 2014: it is not yet perfect, but it is improving all the time. Microsoft CEO Satya Nadella revealed an intriguing discovery which he called transfer learning: 'If you teach it English, it learns English', he said. 'Then you teach it Mandarin: it learns Mandarin, but it also becomes better at English, and quite frankly none of us know exactly why'.[63]

Several companies are proposing to take the obvious logical step, and manufacture earpieces which recognise the wearer's speech, translate it and convey it to a listener. Equally obvious and logical is that these devices will be called Babel Fish, after what Douglas Adams described as the 'oddest thing in the universe' in his 'Hitchhiker's Guide to the Galaxy' series. When that series first appeared on BBC radio in 1978, very few listeners would have believed you if you told them a real version would appear just 40 years later. But now it has.[64] At the moment they are slow, with a couple of seconds delay, and they are error-prone. But those imperfections will doubtless be ironed out within a few years.

Finally, machines can now beat humans at lip-reading, so the scene in the film '2001' where HAL 'eavesdrops' on the astronauts while they discuss his apparent malfunction is another scenario that has stepped out of science fiction and into reality.[65]

2.3.8 Learning and Innovating

It can no longer be said that machines do not learn, or that they cannot invent. In December 2013, DeepMind demonstrated an AI system which used unsupervised learning deep learning to teach itself to play old-style Atari video games like Breakout and Pong.[66] These are games which previous AI systems found hard to play because they involve hand-to-eye coordination.

The system was not given instructions for how to play the games well, or even told the rules and purpose of the games: it was simply rewarded when it played well and not rewarded when it played less well. As the writer Kevin Kelly noted, 'they didn't teach it how to play video games, but how to *learn* to play the games. This is a profound difference'.[67]

The system's first attempt at each game was feeble, but by playing continuously for 24 hours or so it worked out – through trial and error – the subtleties in the gameplay and scoring system, and played the games better than the best human player.

The DeepMind system showed true general learning ability. On seeing the demonstration, Google acquired the company for a reported $400 million.

2.3.9 Creativity

In the last few years, a host of companies have been set up to create music with AI. The CEO of one of them observed in August 2017 that 'a couple of years ago, AI wasn't at the stage where it could write a piece of music good enough for anyone. Now it's good enough for some use cases. It doesn't need to be better than Adele or Ed Sheeran. The aim is not "will this get better than X?" but "will it be useful for people?"[68]

Machines are creating visual imagery as well as music. In June 2015, Google released pictures produced by an image recognition neural network called Deep Dream. They captured the public imagination because of their surreal, hallucinogenic properties. The network was told to look for a particular feature – for instance an eye, or a dog's head – and to modify the picture to emphasise that feature. Repeated iterations around a feedback loop created images that were sometimes beautifully haunting, and sometimes just haunting.[69]

Another creative system, called generative adversarial networks, or GANs, was developed by Google researcher Ian Goodfellow in 2014. The system contains two neural networks – one to generate a candidate image and the other to judge it, or discriminate, based on its database of prior images. Competing in this way, the system is able to produce photo-realistic images which humans cannot detect as fakes.[70]

In December 2015, a group of AI researchers at the Massachusetts Institute of Technology (MIT) published a paper about a system which was able to predict the memorability of images better than humans. The system, called MemNet, reviewed a dataset of 60,000 images, and classified them in 1,000 different ways. It was able to identify why certain images were more memorable than others.

To be clear, these systems are not conscious and have no imagination. They are not emotionally affected by the images they process because they have no emotions. Depending on your definition of art, they may or may not be creating it. I think of art as the use of a craft skill to communicate something profound or at least interesting about the condition of being human. (Admittedly that excludes a lot of what we see in museums and art galleries, but many of us would be OK with that.) Using that definition, machines are not creating art because they have no experience of the condition of being human.

But in some contexts, the conscious experience – or lack of it – behind a creative act does not matter. Machines can analyse, process and even create text, sounds and images in ways that are important to us. And sometimes they can do it better than we can.

2.3.10 Natural Language Processing

ML techniques are enabling machines to parse sentences in ordinary language, reduce them to their basic components, process them in order to achieve a particular result, and offer up the output in a form that humans can easily understand. This is natural language processing, or NLP.

Remember spam? In the late 2000s, there was talk of it crashing the Internet, but now you rarely see it unless you look at your junk mailbox. It was tamed by ML and NLP.

The same is happening with user-generated content (UGC). We like to read the comments on news sites: many of them are dumb, but many are smart and funny. After all, there wouldn't be any point in crowdsourcing if the crowd was all stupid. But some of it needs a grammatical dry-clean to be useful, and the good stuff needs to be surfaced. This is increasingly being done with ML, and by companies far down the pyramid from the tech giants. Companies large and small are using ML to work out what information to present to their customers and targets at every encounter.[71]

Some UGC needs much more than a grammatical dry-clean: some of it needs to be removed. At the time of writing, four hours of video is uploaded to Google's YouTube service every minute, and they would have to employ an implausible number of humans to review it all. Fortunately, Google's AI systems can help: 'Our initial use of machine learning has more than doubled both the number of videos we've removed for violent extremism. ... While these tools aren't perfect, and aren't right for every setting, in many cases our systems have proved more accurate than humans at flagging videos that need to be removed'.[72]

2.3.11 The Science of What We Can't Yet Do

A famous cartoon shows a man in a small room writing notes to stick on the wall behind him. Each note shows an intellectual task which computers are unable to carry out. On the floor, there is a growing pile of discarded notes – notes which show tasks which computers now carry out better than humans. The notes of the floor include 'arithmetical calculations', 'play chess', 'recognise faces', 'compose music in the style of Bach', 'play table tennis'. The notes on the wall include tasks which no computer can do today, including 'demonstrate common sense', but also some tasks which they now can, such as 'drive cars', 'translate speech in real time'.

The man in the cartoon has a nervous expression: he is perturbed by the rising tide of tasks which computers can perform better than humans. Of course, it may be that there will come a time the notes stop moving from the wall to the floor. Perhaps computers will never demonstrate common sense. Perhaps they will never report themselves to be conscious. Perhaps they will never decide to revise their goals. But given their startling progress to date and the weakness of the *a priori* arguments that strong AI cannot be created (which we will review subsequently), it seems unwise to bet too heavily on it.

It is hard to forecast what machines will and will not be able to do, and by when. It turns out to be relatively easy to programme computers to do things that we find very hard, like advanced arithmetic, but very hard to teach them how to do things that we find easy, like tying our shoelaces. This is known as Moravec's paradox, after AI pioneer Hans Moravec.[73]

We tend to forget how much progress AI has made. Although iPhones and Android phones are called 'smartphones', we don't tend to think of them as instantiations of AI. We don't tend to think of the logistical systems of the big supermarkets as examples of AI. In effect, AI is redefined every time a breakthrough is achieved. Computer scientist Larry Tesler pointed out that this means AI is being defined as 'whatever hasn't been done yet', an observation which has become known as Tesler's Theorem, or the AI effect.

For many years, people believed that computers would never beat humans at chess. When it finally happened, it was dismissed as mere computation – mere brute force, and not proper thinking at all. The rebarbative American linguistics professor Noam Chomsky declared that a computer programme beating a human at chess was no more interesting than a bulldozer winning an Olympic gold at weight-lifting. (Maybe he was saying this before Kasparov was beaten, but if so, he was very unusual.)

Computers are not (as far as we know) self-conscious. They cannot reflect rationally on their goals and adjust them. They do not (as far as we know) get excited about the prospect of achieving their goals. They do not (we believe) actually understand what they are doing when they play a game of chess, for instance.

In this sense it is fair to say that what AI systems do is 'mere computation'. Then again, a lot of what the human brain does is 'mere computation', and it has enabled humans to achieve some wondrous things. It is not unreasonable that humans want to preserve some space for themselves at the top of the intellectual tree. But to dismiss everything that AI has achieved as not intelligent, and to conclude – as some people do – that AI research has made no progress since its early days in the 1950s and 1960s is frankly ridiculous.

ML works, and in a few short years it has achieve remarkable things. But in truth, it has hardly got started. In the next chapter, Chapter 3: Exponential Improvement, we will look at what is driving its progress.

NOTES

1. http://www.computerweekly.com/feature/Apollo-11-The-computers-that-put-man-on-the-moon.
2. http://www.bradford-delong.com/2017/09/do-they-really-say-technologi-cal-progress-is-slowing-down.html.
3. http://www.internetlivestats.com/google-search-statistics/.
4. http://en.wikipedia.org/wiki/High-frequency_trading.
5. http://www.strategyand.pwc.com/global/home/what-we-think/innovation1000/top-innovators-spenders#/tab-2015.
6. 2013 data: http://www.ons.gov.uk/ons/rel/rdit1/gross-domestic-expendi-ture-on-research-and-development/2013/stb-gerd-2013.html.
7. http://streamhistory.com/die-rich-die-disgraced-andrew-carnegies-phi-losophy-of-wealth/
8. *The Big Switch* by Nicholas Carr (p. 212).
9. https://www.wired.com/2017/02/inside-facebooks-ai-machine/
10. http://www.theguardian.com/technology/2016/jan/31/viv-artificial-intelligence-wants-to-run-your-life-siri-personal-assistants.
11. https://techcrunch.com/2017/07/19/apple-launches-machine-learning-research-site/.
12. https://techcrunch.com/2016/09/29/microsoft-forms-new-ai-research-group-led-by-harry-shum/.
13. https://www.geekwire.com/2017/one-year-later-microsoft-ai-research-grows-8k-people-massive-bet-artificial-intelligence/.
14. https://techcrunch.com/2017/07/12/microsoft-creates-an-ai-research-lab-to-challenge-google-and-deepmind/.

15. http://uk.businessinsider.com/amazon-echo-sales-figures-stats-chart-2016-12.
16. http://www.aaai.org/Magazine/Watson/watson.php.
17. http://www.latimes.com/business/technology/la-fi-cutting-edge-ibm-20160422-story.html.
18. https://www.top500.org/news/financial-analyst-takes-critical-look-at-ibm-watson/.
19. http://www.cnbc.com/2017/05/08/ibms-watson-is-a-joke-says-social-capital-ceo-palihapitiya.html.
20. https://seekingalpha.com/article/2463495-china-search-engine-market-share-august-2014.
21. http://www.nasdaq.com/article/an-indepth-look-at-baidus-bidu-artificial-intelligence-aspirations-cm821145.
22. http://english.gov.cn/policies/latest_releases/2017/07/20/content_281475742458322.htm.
23. https://topchronicle.com/alibaba-group-holding-limited-baba-has-a-market-value-of-388-89-billion/.
24. https://www.technologyreview.com/s/609099/alibaba-aims-to-master-the-laws-of-ai-and-put-virtual-helpers-everywhere/?utm_campaign=add_this&utm_source=twitter&utm_medium=post.
25. https://www.washingtonpost.com/news/the-switch/wp/2016/10/13/china-has-now-eclipsed-us-in-ai-research/?utm_term=.d4f9eb1c9353 This article is behind a paywall. An alternative source is here: https://futurism.com/china-has-overtaken-the-u-s-in-ai-research/.
26. http://thediplomat.com/2017/07/chinas-artificial-intelligence-revolution/.
27. http://www.shanghaidaily.com/metro/society/Shanghai-will-be-at-the-heart-of-Chinas-artificial-intelligence/shdaily.shtml.
28. http://www.business-standard.com/article/economy-policy/task-force-seeks-public-opinion-on-artificial-intelligence-117092300823_1.html.
29. http://www.wired.com/2015/11/google-open-sources-its-artificial-intelligence-engine/.
30. https://www.theguardian.com/technology/2016/apr/13/google-updates-tensorflow-open-source-artificial-intelligence.
31. http://www.wired.com/2015/12/facebook-open-source-ai-big-sur/.
32. The name Parsey McParseFace is a play on a jokey name for a research ship which received a lot of votes in a poll run by the British government in April 2016.http://www.wsj.com/articles/googles-open-source-parsey-mcparseface-helps-machines-understand-english-1463088180.
33. https://www.forbes.com/sites/bernardmarr/2017/09/18/the-amazing-ways-coca-cola-uses-artificial-intelligence-ai-and-big-data-to-drive-success/#7f7bc55078d2.
34. https://ai-europe.com/ (Disclaimer: I chair this conference.).
35. https://www.satalia.com/.
36. http://rainbird.ai/.
37. https://www.crowdflower.com/.
38. http://insights.venturescanner.com/category/artificial-intelligence-2/.

39. http://techcrunch.com/2015/12/25/investing-in-artificial-intelligence/.

40. https://www.youtube.com/watch?v=Skfw282fJak.

41. http://futureoflife.org/2016/01/27/are-humans-dethroned-in-go-ai-experts-weigh-in/.

42. https://www.theverge.com/2017/8/11/16137388/dota-2-dendi-open-ai-elon-musk.

43. http://www.popsci.com/scitech/article/2004-06/darpa-grand-challenge-2004darpas-debacle-desert.

44. https://www.theguardian.com/technology/2016/mar/09/google-self-driving-car-crash-video-accident-bus.

45. http://www.thechurchofgoogle.org/Scripture/Proof_Google_Is_God.html.

46. https://www.reddit.com/r/churchofgoogle/.

47. https://www.theguardian.com/technology/2017/sep/10/are-flying-cars-ready-for-takeoff.

48. The answer, if you're searching from England, is to fly over Asia.

49. http://searchengineland.com/faq-all-about-the-new-google-rank-brain-algorithm-234440?utm_campaign=socialflow&utm_source=facebook&utm_medium=social.

50. http://www.wired.com/2016/02/ai-is-changing-the-technology-behind-google-searches/.

51. http://www.thedrum.com/opinion/2016/02/08/why-artificial-intelligence-key-google-s-battle-amazon.

52. http://www.wired.com/2012/06/google-x-neural-network/.

53. They are the Pembroke and the Cardigan Corgi. http://research.microsoft.com/en-us/news/features/dnnvision-071414.aspx.

54. http://image-net.org/challenges/LSVRC/2015/index#news.

55. http://www.eetimes.com/document.asp?doc_id=1325712.

56. https://youtu.be/U_Wgc1JOsBk?t=33.

57. http://www.wired.com/2016/01/2015-was-the-year-ai-finally-entered-the-everyday-world/.

58. At the time of writing, April 2016, Aipoly is impressive, but far from perfect.

59. http://news.sciencemag.org/social-sciences/2015/02/facebook-will-soon-be-able-id-you-any-photo.

60. http://www.computerworld.com/article/2941415/data-privacy/is-facial-recognition-a-threat-on-facebook-and-google.html.

61. http://www.bloomberg.com/news/2014-12-23/speech-recognition-better-than-a-human-s-exists-you-just-can-t-use-it-yet.html.

62. http://www.businessinsider.fr/uk/microsofts-speech-recognition-5-1-error-rate-human-level-accuracy-2017-8/.

63. http://www.forbes.com/sites/parmyolson/2014/05/28/microsoft-unveils-near-real-time-language-translation-for-skype/.

64. https://www.theguardian.com/technology/2017/oct/05/google-pixel-buds-babel-fish-translation-in-ear-ai-wireless-language.

65. http://www.zdnet.com/article/lip-reading-bots-in-the-wild/.

66. https://youtu.be/V1eYniJ0Rnk?t=1.

67. http://edge.org/response-detail/26780.

68. https://www.theguardian.com/technology/2017/aug/06/artificial-intelligence-and-will-we-be-slaves-to-the-algorithm.

69. https://www.theguardian.com/technology/2015/jun/18/google-image-recognition-neural-network-androids-dream-electric-sheep.

70. https://www.youtube.com/watch?v=deyOX6Mt_As

71. http://techcrunch.com/2016/03/19/how-real-businesses-are-using-machine-learning/

72. https://www.theguardian.com/technology/2017/aug/01/google-says-ai-better-than-humans-at-scrubbing-extremist-youtube-content?utm_source=esp&utm_medium=Email&utm_campaign=GU+Today+main+NEW+H+categories&utm_term=237572&subid=17535328&CMP=EMCNEWEML6619I2.

73. Moravec wrote about this phenomenon in his 1988 book *Mind Children*. A possible explanation is that the sensory motor skills and spatial awareness that we develop as children are the product of millions of years of evolution. Rational thought is something we have only been doing for a few thousand years. Perhaps it really isn't hard, but just seems hard because we are not yet optimised for it.

CHAPTER 3

Exponential Improvement

3.1 EXPONENTIALS

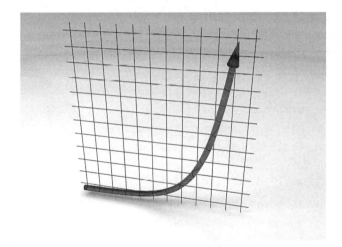

3.1.1 The Power of Exponential Growth

Using a car as a metaphor for artificial intelligence (AI), algorithms are the engine control system, big data is the fuel and computing power is the engine. The engine is getting more powerful at an exponential rate: its performance is doubling repeatedly. It is impossible to understand the scale of change that we face in the coming years without comprehending the astonishing impact of exponential increase.

Imagine that you stand up and take 30 paces forward. You would travel around 30 metres. Now imagine that you take 30 exponential paces, doubling the length each time. Your first pace is one metre, your second is two metres, your third is four metres, your fourth pace is eight metres, and so on.

How far do you think you would travel in 30 paces? The answer is, to the moon. In fact, to be precise, the 29th pace would take you to the moon; the 30th pace would bring you all the way back.

That example illustrates not just the power of exponential increase, but also the fact that it is deceptive, and back-loaded.

Here is another illustration. Imagine that you are in a football stadium (either soccer or American football will do) which has been sealed to make it waterproof. The referee places a single drop of water in the middle of the pitch. One minute later she places two drops there. Another minute later, four drops, and so on. How long do you think it would take to fill the stadium with water? The answer is 49 minutes. But what is really surprising – and disturbing – is that after 45 minutes, the stadium is just 7% full. The people in the back seats are looking down and pointing out to each other that something significant is happening. Four minutes later they have drowned.[1]

The fact that exponential growth is back-loaded helps explain another phenomenon, known as Amara's Law, after the scientist Roy Amara. This states that we tend to overestimate the effect of a technology in the short run and underestimate the effect in the long run.[2]

3.1.2 No Knees

People often talk about the 'knee' of an exponential curve, the point at which past progress seems sluggish, and projected future growth looks dramatic. This is a misapprehension. When you compare exponential curves plotted for ten and 100 periods of the same growth, they look pretty much the same. In other words, wherever you are on the curve, you are always at the beginning of the real growth phase: the past always looks horizontal and the future always looks vertiginous.

The author John Lanchester describes how in 1996 the US government started building a new supercomputer to model the behaviour of nuclear explosions. The result was Red, the first machine to process more than a trillion floating point operations per second (a teraflop). It remained the world's fastest supercomputer until 2000, but by 2006 that level of processing was available to schoolchildren in the Sony PS3 gaming computer.[3] This is Moore's law at work.

3.2 MOORE'S LAW

In 1965, Gordon Moore was working for Fairchild Semiconductors. He published a paper observing that the number of transistors being placed on a chip was doubling every year. He forecast that this would continue for a decade, which his contemporaries considered extremely adventurous. In 1975, he adjusted the period to 2 years, and shortly afterwards, a CalTech professor named Carver Mead coined the term Moore's law. In 1968, Moore co-founded Intel, and following an observation by Intel executive David House that the performance of individual transistors was also improving, the Law is generally taken to mean that the processing power of $1,000-worth of computer doubles every 18 months.

Moore's law is not a law, but an observation which became a self-fulfilling prophecy – a target and a planning guide for the semiconductor industry, and for Intel in particular. Moore's law celebrated its 50th anniversary in April 2015. Given its importance in human affairs, there was remarkably little fanfare.

Exponential curves do not generally last for long: they are just too powerful. In most contexts, fast-growing phenomena start off slowly, pick up speed to an exponential rate, and then after a few periods they tail off to form an S-shaped curve. However, exponentials can continue for many steps, and in fact each of us is one of them. You are composed of around 27 trillion cells, which were created by fission, or division – an exponential process. It required 46 steps of fission to create all of your cells. Moore's law, by comparison, has had 33 steps in the 50 years of its existence.

It would take another two decades for Moore's law to run through the same number of steps as human cell division. By that time, if the Law holds, the total amount of computing power now available to Google will be available on an ordinary desktop computer. Imagine what we could achieve when every teenager has Google's computing power in her bedroom – and try to imagine how much computing power Google will have by then!

A number of other technological developments have been observed to progress at an exponential rate, including memory capacity, LEDs (which follow Haitz's Law[4]), sensors (where the cost per observation is decreasing exponentially[5]) and the number of pixels in digital cameras.[6]

3.2.1 No More Moore?

In 2015, Intel seemed uncertain about whether its own chip development would keep Moore's law on track. This is important, as Intel (whose name is a compression of Integrated Electronics) has been the world's biggest chip manufacturer since 1991, and is also the world's third-largest R&D spending company (after Volkswagen and Samsung). It has led the miniaturisation of microchips.

In February 2015, Intel updated journalists on their chip programme for the next few years, and the schedule maintained Moore's law's exponential growth.[7] The first chips based on its new 10-nanometre manufacturing process were due to be released in late 2016/early 2017, after which Intel expected to move away from silicon, probably towards a III-V semiconductor such as indium gallium arsenide.[8] (That 10 nanometres is the distance between the two nearest repeating features on the chip.)

But in July 2015, Intel CEO Brian Krzanich said that it was taking longer for the firm to cut the size of its transistors: 'our cadence today is closer to 2.5 years than to 2'. At the time of writing, the firm's smallest transistors are the 14-nanometre Skylake model, and the next size down will be the 10-nanometre Cannonlake, due in late 2017, a 6-month delay.

Since 2007, Intel had pursued the development of its chips with a 'tick-tock' cadence. The tick represented improvements in the manufacturing process, which enabled chip size to be reduced from 45 to 32 to 22 to 14 nanometres. The tock was improvements in the architecture. The new cadence announced by Krzanich was described by one observer as a move from tick-tock to tic-tac-toe, representing process (tic), architecture (tac) and optimisation and efficiency (toe).[9]

However, Mark Bohr, a 37-year Intel veteran and senior fellow in its processor technology team, argued that taking a longer view, Moore's law still applied. He is working on the technology to get down to five nanometres.[10] And in January 2017, Krzanich unveiled a laptop containing the Canonlake chip, saying 'I've heard the death of Moore's law more times than anything else in my career. And I'm here today to really show you and tell you that Moore's law is alive and well and flourishing'.[11]

(To provide some context, a human hair is 100,000 nanometres thick, so each hair on your head is 10,000 times thicker than Intel's next release of transistors. Silicon atoms are around 0.2 nanometres across, so a 5-nanometre structure is about 20-atoms wide.)

Intel is the largest chip maker, but it is not the only game in town. In July 2015, IBM announced that it had a prototype chip of seven nanometres, using silicon-germanium for some components.[12] Manufacturing at scale is very different from prototyping, however, and IBM did not expect to manufacture at scale for two more years. But this would be ahead of Intel's schedule, and the IBM announcement was especially important in heralding a successful move from Deep UltraViolet (DUV) to Extreme UltraViolet (EUV) lithography, which operates at much shorter wavelengths.

3.2.2 New Architectures

Moore's law has undergone substantial transitions before. Until 2004, regular increases in the clock speeds of computer chips contributed a large part of their performance improvements (see Note 13 for an explanation of clock speeds, if you like that kind of thing.)[13] Overheating put a stop to this, and instead, chip manufacturers incorporated more processors, or 'cores'. Modern smartphones may have as many as eight, which means the processes they work on have to be broken down into pieces which are operated on in parallel.

However successfully the chip manufacturers prolong it, the existing architecture will reach its endpoint eventually. Researchers are hard at work on a number of technologies that could prolong or replace it.

Intel based its dominance on central processing units, or CPUs. They are general-purpose processors which can carry out many kinds of computation, but are not necessarily optimised for any of them. The requirements of video games prompted the development and refinement of graphics processing units, or GPUs. These are very good at taking huge quantities of data and carrying out the same operation over and over again. It turns out that machine learning (ML) benefits from their particular capabilities. CPUs and GPUs are often deployed in tandem.

The champion of GPUs is Nvidia, whose value rose ten-fold in 3 years to 2017, to more than $100 billion, as its chips have proved invaluable for manufacturers of mobile phones as well as games consoles. In 2015 and 2016, Intel spent $17 billion on two companies, Altera and Nervana, in order to keep catch up with Nvidia.

Still more specialised than GPUs are Google's TPUs, or tensor processing units. These operate at very high speeds with lower precision than GPUs and CPUs, and are specifically designed for ML applications.[14]

Another new architecture being developed is 3D chips. Placing chips side-by-side delays the signals between them and causes bottlenecks as too many signals try to use the same pathways. These problems can be eased if you place the chips on top of each other, but this raises new problems. Silicon chips are fabricated at 1800°F, so if you manufacture one chip on top of another you will fry the one below. If you fabricate them separately and then place one on top of the other, you have to connect them with thousands of tiny wires.

In December 2015, researchers from Stanford announced a new method of stacking chips which they called nanoengineered computing systems technology, or N3XT. They claimed this was a thousand times more efficient than conventional chip configurations.[15] They did not give an estimate when mass production of N3XT chips might begin.

Another approach is to combine the memory chips with the traditionally separate processing chips, to reduce the amount of traffic between those two. Another is to design chips specifically to implement neural networks, which is the approach taken by a Massachusetts Institute of Technology (MIT) team that announced the Eyeriss chip in February 2016.[16]

In March 2016, scientists from IBM's TJ Watson Research Center announced their belief that 'resistive processing units' which combine CPU and memory on the same chip could accelerate the processing of ML algorithms as much as 30,000 times.[17]

Carver Mead is a pioneer of the microelectronics industry based at CalTech; he gave Moore's law its name. In the late 1980s, he proposed the idea of computers based on the architecture of the brain, called neuromorphic computing. As with many developments in computing, it could not be realised at the time, but is now becoming a reality. In 2014, IBM produced a chip called TrueNorth, which comprised around a million silicon 'neurons', each with 256 'synapses'. As well as being very powerful, IBM claimed the chip is extremely energy-efficient.

As the world's largest chip manufacturer, Intel is also working on neuromorphic computing, and in September 2017, it unveiled Loihi, which it described as the world's first self-learning neuromorphic chip: it mimics how the brain functions by learning from feedback received from its environment. As well as being very energy-efficient, this type of chip is expected to be a very effective platform for deep learning systems. The Loihi chip will be shared with leading AI research labs from early 2018.[18]

3.2.3 Towards Quantum Computing

Quantum computing, like nuclear fusion, is one of those technologies which seems forever to be 20 years away from any practical application, but experienced researchers are now saying its time is coming soon. It is based on the idea that while classical computers use bits (binary digits) which are either on or off, quantum bits (qubits) can be both on and off at the same time – known as superposition. This enables them to carry out a number of different calculations at once.

Google bought a quantum computer from Canadian company D-Wave in 2013, but was unable to demonstrate to everyone's satisfaction that it actually worked until December 2015, when Google's engineering director Hartmut Neven announced that its D-Wave computer was 100 million times faster than a traditional desktop computer in a 'carefully crafted proof-of-concept problem'. There is still debate about whether D-Wave's so-called 'quantum annealing computers' are true quantum computers at all, but in any case, general-purpose quantum computers are still under development.

Keeping qubits stable is very hard, but Google, IBM and Microsoft all think they are getting close to 'quantum supremacy', the point where quantum computers are more powerful than classical ones – at least for some types of calculations. Fifty qubits is generally thought to be the number needed for quantum supremacy. IBM currently offers researchers the opportunity to perform calculations on a basic quantum computer with 16 qubits.

In November 2016, Microsoft hired four top quantum computing researchers, and their leader claimed that 'we are ready to go from research to engineering'. The following September, the company announced a multimillion-dollar investment to build their quantum research team in collaboration with Copenhagen University.[19,20]

In mid-2017, Google reported that it was testing a 20-qubit device, and expected to have a prototype with 49 qubits by the end of the year.[21] It was beaten to the punch by a team of scientists from Russia and from Harvard University in the United States, who announced the successful test of a 51-qubit device in August.[22]

3.3 SO WHAT?

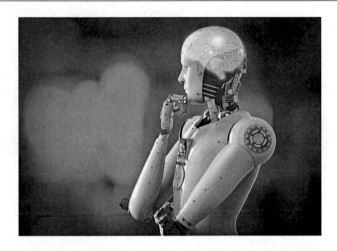

Moore's law was an observation which became a target-generator rather than being a description of a fundamental property of the world. Keeping it on track has involved numerous ingenious and unpredictable steps in the past. Commercial imperatives and sheer human inventiveness have managed it so far, and there are plenty of avenues being explored which could maintain that performance.

Using the specific definition arrived at by Gordon Moore and David House in the late 1960s, Moore's law ceased to hold a long time ago – a fact which some people delight in pointing out, although they rarely explain which of the numerous definitions they are referring to – Moore's original one, his revised one, David House's one, or something else.

As Intel's head of advanced microprocessor research Shekhar Borkar observed, from the consumer perspective, 'Moore's law simply states that user value doubles every 2 years'. There are plenty of ways to keep that going, and plenty of incentives too.[23] Perhaps those who love to claim that Moore's law is dead should invent some other name for what the rest of us are using it to mean – although it probably wouldn't catch on.

The people who are actually working on the technologies seem determined to maintain the pace of the advance. Moore is more, and more is better. The consensus is clear: there will be continued rapid improvement in computer processing power.

3.3.1 Machines with Common Sense?

In December 2015, Microsoft's chief speech scientist Xuedong Huang noted that speech recognition has improved 20% a year consistently for the last 20 years. He predicted that computers would be as good as humans at understanding human speech by 2021. Geoff Hinton – the man whose team won the landmark 2012 ImageNet competition – went further. In May 2015, he said that he expects machines to demonstrate common sense within a decade.[24]

Common sense can be described as having a mental model of the world which allows you to predict what will happen if certain actions are taken. Professor Murray Shanahan of Imperial College uses the example of throwing a chair from a stage into an audience: humans would understand that members of the audience would throw up their hands to protect themselves, but some damage would probably be caused, and certainly some upset. A machine without common sense would have very little idea of what would happen.

Facebook has declared its ambition to make Hinton's prediction come true. To this end, it established a basic research unit in 2013 called Facebook AI Research (FAIR) with 50 employees, separate from the 100 people in its applied machine learning team.[25]

Many other researchers are working in the same direction. An MIT professor called Josh Tenenbaum was part of a team which in 2015 demonstrated an AI programme that was able to recognise a handwritten character based on just one example, as opposed to the millions of data points required by most ML systems.[26] In 2017, he was applying the approach to a Boston-based, self-driving car start-up called iSee.[27]

So within a decade, machines will be better than humans at recognising faces and other images, better at understanding and responding to human speech, and may even be possessed of common sense. And they will be getting faster and cheaper all the time. This phenomenon will have such enormous impact on our lives in the years and decades ahead that it is surprising how little attention it receives from policy makers and their advisers. It did not feature significantly at all in any of the recent elections held in Europe or the United States, for instance.

3.3.2 Why Don't We Notice the Exponential Rate of Improvement in AI?

Most of the things which are of vital interest to us change at a linear rate. Leopards chase us and our prey eludes us one step at a time; the seasons change one day at a time. We have a hard time adjusting to things that change at an exponential rate. But even when it is pointed out to us, we often choose to ignore it. Here are six possible reasons why many people shrug off the extraordinary progress of AI as not significant.

1. In the last chapter, we encountered Tesler's Theorem, in which AI is defined as 'whatever hasn't been done yet'. It is easy to dismiss AI as not progressing very fast if you allocate all its achievements to some other domain.

2. The demise of Moore's law has been predicted ever since it was devised 50 years ago. There is no need to worry about a phenomenon which is almost over.

3. We noted in the last chapter that we are curiously nostalgic about the future that we once thought we would have. We haven't got hoverboards, flying cars or personal jet packs, but we have got pretty close to omniscience at the touch of a button. (And we may soon get flying cars too.) What's coming next is no less amazing, but we tend to focus on what we didn't get more than what we did.

4. The early incarnations of new inventions are often disappointing. If you were around for the launch of the first mobile phones you will remember they were the size of a brick and the weight of a small suitcase; they were ridiculed as the expensive playthings of pretentious yuppies. Now almost everyone in the developed world has a smartphone. Similar ridicule attended the launch of Siri and Google Glass, but contrary to popular opinion, they are emphatically not failures. They are simply the first, tentative outings of technologies which will soon revolutionise our lives. We forget the often-repeated lesson: the technology will improve, and its contribution to our productivity and our effectiveness will be substantial.

5. In fact, these technologies are simply following the standard curve of the product life cycle. At the initial launch of a new product, a small tribe of what marketers call 'innovators' jump on it because it

is new and shiny. They can see its potential and they generate some early hype. Some 'early adopters' then try it out and declare it not fit for purpose – and they are right. The backlash sets in, and a wave of cynicism submerges all interest in the product. Over successive months or years, the technology gradually improves, and eventually crosses a threshold at which point it is fit for purpose. In technology marketing circles this is known as crossing the chasm, and of course many technologies never manage it. They never find their killer application.

Those technologies which do cross the chasm are then embraced by the early adopters and the early majority, then the late majority, and finally by the laggards. But by the time the early majority is getting on board, the hype is already ancient history, and people are already taking for granted the improvement to their lives. The hype cycle has run its course.

6. The hedonic treadmill is a name for the fact that most people have a fairly constant level of happiness (hedonic level), and that when something significant in our life changes – for good or bad – we quickly adjust and return to our previous level. When we look ahead to an anticipated event, we often believe that it will change our lives permanently, and that we will feel happier – or less happy – forever afterwards. When the event actually happens, we quickly become accustomed to the new reality, and what seemed wonderful in prospects becomes ordinary. 'Wow' quickly becomes 'meh'.

7. Learning about a new AI breakthrough can be slightly unsettling. There is a vague awareness that we may be creating our own competition, and of course there is the image of the Terminator, which has been entertaining us frightfully for over 30 years. At a deep level, perhaps many of us really don't want to acknowledge the exponential rate of AI improvement.

In the next chapter, Chapter 4: Tomorrow's AI, we will try to suppress this latent fear, and explore what AI will bring us in the next few years.

NOTES

1. I am grateful to Russell Buckley for introducing me to this illustration.
2. http://www.pcmag.com/encyclopedia/term/37701/amara-s-law.

3. http://www.lrb.co.uk/v37/n05/john-lanchester/the-robots-are-coming.
4. Haitz's Law states that the cost per unit of useful light emitted decreases exponentially.
5. http://computationalimagination.com/article_cpo_decreasing.php.
6. http://www.nytimes.com/2006/06/07/technology/circuits/07essay.html.
7. http://arstechnica.com/gadgets/2015/02/intel-forges-ahead-to-10nm-will-move-away-from-silicon-at-7nm/.
8. The 'III-V' refers to the periodic table group the material belongs to. Transistors made from these semiconductors should consume far less power, and also switch much faster.
9. http://www.extremetech.com/extreme/225353-intel-formally-kills-its-tick-tock-approach-to-processor-development.
10. http://www.nextplatform.com/2015/11/26/intel-supercomputer-powers-moores-law-life-support/.
11. http://fortune.com/2017/01/05/intel-ces-2017-moore-law/.
12. http://www.theguardian.com/technology/2015/jul/09/moores-law-new-chips-ibm-7nm.
13. Clock speed, also known as clock rate or processor speed, is the number of cycles a chip (CPU) performs each second. Inside each chip is a small quartz crystal which vibrates, or oscillates, at a particular frequency. It takes a fixed number of oscillations, or cycles, to perform the instructions that a chip is given. One cycle is one Hertz, and today's chips operate in gigaHertz (GHz), billions of cycles per second. As other aspects of chip designs diverge, clock speed is no longer a reliable measure of a chip's effective performance.
14. https://techcrunch.com/2017/05/17/google-announces-second-generation-of-tensor-processing-unit-chips/.
15. http://www.popularmechanics.com/technology/a18493/stanford-3d-computer-chip-improves-performance/.
16. http://gadgets.ndtv.com/science/news/mit-builds-low-power-artificial-intelligence-chip-for-smartphones-799803
17. http://www.engadget.com/2016/03/28/ibm-resistive-processing-deep-learning/.
18. https://newsroom.intel.com/editorials/intels-new-self-learning-chip-promises-accelerate-artificial-intelligence/.
19. http://www.zdnet.com/article/microsofts-next-big-bet-clue-its-just-hired-four-top-quantum-computing-scientists/#ftag=CAD-00-10aag7e.
20. http://www.zdnet.com/google-amp/article/microsoft-just-upped-its-multi-million-bet-on-quantum-computing/.
21. http://uk.businessinsider.com/google-quantum-computing-chip-ibm-2017-6?r=US&IR=T.
22. https://www.sciencealert.com/google-s-quantum-announcement-over-shadowed-by-something-even-bigger.
23. http://www.nature.com/news/the-chips-are-down-for-moore-s-law-1.19338.
24. https://www.theguardian.com/science/2015/may/21/google-a-step-closer-to-developing-machines-with-human-like-intelligence.

25. http://fortune.com/facebook-machine-learning/.
26. https://www.technologyreview.com/s/544376/this-ai-algorithm-learns-simple-tasks-as-fast-as-we-do/.
27. https://www.technologyreview.com/s/608871/finally-a-driverless-car-with-some-common-sense/?utm_campaign=add_this&utm_source=twitter&utm_medium=post.

CHAPTER 4

Tomorrow's AI

4.1 A DAY IN THE LIFE

Julia woke up feeling rested and refreshed. This was unremarkable: she had done so ever since Hermione, her digital assistant, had been upgraded to monitor her sleep patterns and wake her at the best stage in her sleep cycle. But Julia could still remember what it was like to wake up to the sound of an alarm crashing through REM sleep and she felt grateful as she stretched her arms and smelled the coffee which Hermione had prepared.

Traffic feeds indicated that the roads were quiet, so with a few minutes in hand, Hermione updated her on her key health indicators – blood pressure, cholesterol levels, percentage body fat, insulin levels and the rest. Julia had long since stopped feeling disturbed by the idea of the tiny monitors that nestled in all parts of her body, including her bloodstream, her eye fluids, her internal organs and her mouth.

She slipped into the outfit that she and Hermione had put together last night with the aid of some virtual wardrobe browsing and a bit of online shopping. An airborne drone had dropped off the new dress while Julia was asleep; selecting the right size was rarely a problem now that virtual mannequins were supported by most retailers. The finishing touch was an intriguing necklace which Julia had her 3D printer produce overnight in a colour which complemented the dress perfectly, based on a design sent down by her sister in Edinburgh.

During the drive to the station, Julia read a couple of the items which Hermione had flagged for her in the morning's newsfeed, along with some gossip about a friend who had recently relocated to California. With a gesture inside the holosphere, projected by Hermione for the purpose, Julia OK'd the micro-payments to the newsfeeds. Half an hour later, she smiled to herself as her self-driving car slotted itself perfectly into one of the tight spaces in the station car park. It was a while since she had attempted the manoeuvre herself; she knew she would not have executed it so smoothly even when she used to do her own driving. She certainly wouldn't be able to do it now that she was so out of practice.

The train arrived soon after she reached the platform (perfect timing by her car, again) and Hermione used the display in Julia's augmented reality (AR) contact lenses to highlight the carriage with the most empty seats, drawing on information from sensors inside the train. As Julia boarded the carriage, the display highlighted the best seat to choose, based on her travelling preferences and the convenience of disembarkation at the other end of the journey.

Julia noticed that most of her fellow passengers wore opaque goggles: They were watching entertainments with fully immersive virtual reality (VR) sets. She didn't join them. She was going to be spending at least a couple of hours in full VR during the meeting today and she had a personal rule to limit the numbers of hours spent in VR each day.

Instead, she kept her AR lenses in and gazed out of the window. The train took her through parts of the English countryside where she could choose between overlays from several different historical periods. Today she chose the Victorian era, and enjoyed watching how the railway she was travelling was under construction in some parts, with gangs of labourers laying down tracks. She marvelled at that kind of work being done by humans rather than machines.

Hermione interrupted her reverie, reminding Julia that tomorrow was her mother's birthday. Hermione displayed a list of suggested presents for Julia to order for delivery later today, along with a list of the presents she had sent in recent years to make sure there was no tactless duplication. Julia chose the second gift on the list and authorised payment along with Hermione's suggested greeting.

That done, Hermione left her to indulge in her historical browsing for the rest of the journey. The augmented views became increasingly interesting as the train reached the city outskirts, and huge Victorian construction projects unfolded across south London. Julia noticed that the content makers had improved the views since the last time she watched it, adding a host of new characters, and a great deal more information about the buildings in the accompanying virtual menus.

When she reached her office, she had a half hour to spare before the meeting started, so she crafted an introductory message to an important potential new client. She used the latest psychological evaluation algorithm to analyse all the target's publicly available statements, including blog posts, emails, comments and tweets. After reading the resulting profile, she uploaded it into Hermione to help with the drafting. Hermione suggested various phrases and constructions which helped Julia to keep the message formal, avoiding metaphors and any kind of emotive language. The profile suggested that the target liked all claims to be supported by evidence, and didn't mind receiving long messages as long as they were relevant and to the point.

It was time for the conference call – the main reason she had come into the office today. She was meeting several of her colleagues in VR because they were based in ten different locations around the world, and the topic was important and sensitive, so they wanted the communication to be as rich as possible, with all of them being able to see each other's facial expressions and body language in detail. Her VR rig at home wasn't sophisticated enough to participate in this sort of call.

A competitor had just launched a completely automated version of one of Julia's company's major service lines in two countries, and it would probably be rolled out worldwide within a couple of weeks. Julia and her colleagues had to decide whether to abandon the service line, or make the necessary investment to follow suit in automating it – which meant retraining a hundred members of staff, or letting them go.

As always, Julia was grateful to Hermione for discreetly reminding her of the personal details of the colleagues in the meeting that she knew least well. In the small talk at the start and the end of the call, she was able to enquire about their partners and children by name. It didn't bother her at all that their ability to do the same was probably thanks at least in part to their own digital assistants.

Several of the participants in the call did not speak English as their first language, so their words were translated by a real-time machine translation system. The words Julia heard them speak did not exactly match the movement of their mouths, but the system achieved a very believable representation of their vocal characteristics and their inflections.

A couple of times during the call, Hermione advised Julia to slow down, or get to the point faster, using the same psychological evaluation software which had helped to craft the sales message earlier, and also using body language evaluation software.

After the meeting Julia had lunch with a friend who worked in a nearby office. Hermione advised her against the crème brûlée because it would take her beyond her target sugar intake for the day. Julia took a little guilty pleasure in ignoring the advice, and she was sure the dessert tasted better as a result. Hermione made no comment, but adjusted Julia's target for mild aerobic exercise for the day, and set a reminder to recommend a slightly longer time brushing and flossing before bed.

Before heading back to the office, Julia and her friend went shopping for shoes. She was about to buy a pair of killer heels when Hermione advised her that the manufacturer was on an ethical blacklist that Julia subscribed to.

Back in the office, Julia set about reconfiguring the settings on one of the company's lead generation websites. The site used evolutionary algorithms which continuously tested the impact of tiny changes in language, colour and layout, adjusting the site every few seconds to optimise its performance according to the results. This was a never-ending process because the Web itself was changing all the time, and a site which was perfectly optimised at one moment would be outmoded within minutes unless it was re-updated. The Internet was a relentlessly Darwinian environment, where the survival of the fittest demanded constant vigilance.

Reconfiguring the site's settings was a delicate affair because any small error would be compounded rapidly by the evolutionary algorithm, which could quickly lead to unfortunate consequences. She decided to do it with a sophisticated new type of software she had read about the day before. The software provider had supplied detailed instructions to follow with

AR glasses. She went through the sign-up, login and operating sequences carefully, comparing the results with the images shown in her glasses.

By the time Julia was on the train home it was getting dark. She asked Hermione to review the recording of her day and download the new app her friend had recommended in passing at lunchtime. Then she queued up two short videos and four new songs for the drive from the station to her house. She was the only person to get off the train at her home station, so she asked Hermione to display the camera feeds on the two routes she could take from the platform to her car. There was no one on the bridge over the tracks, but there was someone loitering in a shadowy part of the underpass. She took the bridge, even though it was a slightly longer route.

When she reached her car, she checked that the delivery drone had opened her car's boot successfully and dropped the groceries there before relocking it with the remote locking code. Happy in the full possession of her vegetables, she drove home, humming along to Joni Mitchell.

4.2 THE NEW ELECTRICITY

The science fiction writer William Gibson is reported as saying that 'The future is already here – it's just not evenly distributed'.[1] Most of the things mentioned in the previous short story are already available in prototype and early incarnations, and most of the rest is under development. It could take anywhere from 5 to 30 years for you to have working versions of all of them.

Some people will think the life described above is frightening, perhaps dehumanised. It is likely that more people will welcome the assistance, and of course generations to come will simply take it for granted. As Douglas Adams said, anything that is in the world when you're born is just a natural part of the way the world works, anything invented between when you're 15 and 35 is new and exciting, and anything invented after you're 35 is against the natural order of things, and should be banned.[2]

Of course, there is no guarantee that the future will work out this way – in fact, the details are bound to be different. For example, we don't yet know whether the myriad devices connecting up to the Internet of things (IoT) will communicate with us directly, or via personal digital assistants like Hermione. Will you be reminded to take your pill in the morning because its bottle starts glowing, or will your version of Hermione alert you? No doubt the outcome will seem obvious in hindsight.

It has been said that all industries are now part of the information industry – or heading that way. Much of the cost of developing a modern car – and much of the quality of its performance – lies in the software that controls it.

Demis Hassabis has said that AI converts information into knowledge, which he sees as empowering people. The mission statement of Google, which owns his company, is to organise the world's information and make it universally accessible and useful. For many of us, most of the tasks that we perform each day can be broken down into four fundamental skills: looking, reading, writing and integrating knowledge. AI is already help-ing with all these tasks in a wide range of situations, and its usefulness is spreading and deepening.

Marketers used to observe that much of the value of a product lies in the branding – the emotional associations which surrounds it. The same is now true of the information which surrounds it. You might think that the commercial success of a product as physical as say, skin cream does not rely on the provision of information to its consumers. Increasingly that is wrong. Consumers have access to staggering amounts of information about skincare, and many of them want to know how each product they might use would affect their overall regime. In a world of savvy consum-ers, the manufacturer which provides the most concise, easy-to-navigate advice is going to win market share.

From supermarket supply chains to consumer goods to construction to exploring for minerals and oil, the ability to crunch bigger and bigger data sets and make sense of them is improving pretty much every type of human endeavour. Kevin Kelly, the founder of *Wired* magazine, said the business plans of the next 10,000 start-ups are easy to predict: 'Take X and add AI'.[3] To coin a phrase, blessed are the geeks, for they shall inherit the Earth.

Andrew Ng, formerly of Google and Baidu, likes to say that AI is the new electricity. It is phenomenally powerful, increasingly ubiquitous and it powers more and more of the things we rely on in our everyday lives. We will increasingly take it for granted, indeed ignore it, but if it was taken away we would howl with protest. It will also change everything. As Ng says: 'Just as electricity transformed almost everything 100 years ago, today I actually have a hard time thinking of an industry that I don't think AI will transform in the next several years'.[4]

Of course, electricity is also a highly regulated industry which raises great concern about the emissions which are produced while generating it. Parts of the electricity generation and distribution industries are natural monopolies, and other parts are carved up by oligopolistic companies which are rarely much loved. Silver linings are usually accompanied by clouds.

4.3 NEAR-TERM AI

4.3.1 Voice Control

In the coming years, the medium though which we humans communicate with machines will increasingly be voice rather than the combination of screen plus mouse and keyboard. This will make computers easier and faster to deal with, and will broaden the range of situations in which they are useful.

It will also be helpful because many of the people who currently have no Internet access and will be coming online in the next decade do not type. Typing is not a skill typically acquired by people living on a few dollars a day and unable to read.

Voice control is made possible by natural language processing, which we discussed in Section 2.3.10. It enables AI systems to ingest information, process it and return it with added value, all in natural language. At the moment, this takes large amounts of processing power and considerable expertise, but the systems are getting more capable and more streamlined all the time. Two of the biggest applications are chatbots and digital assistants.

Chatbots enable organisations to provide useful information to their customers and stakeholders without the expense of employing lots of humans in call centres. They are a major focus of AI development work within businesses today.[5] You may well have had an exchange with a machine on a company website already, thinking it was a human. This experience will become more common. But the really big application of voice control will be our digital assistants. At the moment, Siri is a bit of a joke. But Siri is going to become serious.

4.3.2 Digital Assistants

We saw in Chapter 2: The State of the Art, that the tech giants are competing to launch the best voice-activated digital assistants. The team developing Amazon's Alexa is ambitious: 'The thing I am sure of is that this time next year she will be significantly more intelligent than she is now, and that sometime in the future we will hit our goal of reinventing the Star Trek computer'.[6] An important part of voice control is that the machine must understand the context of a conversation and be able to answer follow-up conversations. This requires the machine to know what pronouns like 'this', 'that', 'then' and 'her' are referring to.

Alexa and her rivals are primitive today, but in a decade or two their descendants will be our constant companions, and we will wonder how we ever got along without them. They will be our gateway to the Internet, and our invaluable assistants as we navigate our way through the world. Among other things, they will negotiate with and filter out most of the IoT. Although we may not notice it, this will be a blessed relief. Imagine living in a world where every AI-enabled device has direct access to you, with every chair and handrail pitching their virtues to you, and every shop screaming at you to buy something. This dystopia was captured in the famous shopping mall scene in the 2002 film 'Minority Report', and more laconically in Douglas Adam's peerless *Hitchhiker's Guide to the Galaxy* series, where the 'Corporation' that produces the eponymous guide has installed talking lifts, known as happy vertical people transporters. They are extremely irritating.

4.3.3 Friends?

What generic name will be adopted for these assistants? Most of the essential tools which we use every day have one-syllable names, like phone, car, boat, bike, plane, chair, stove, fridge, bed, gun.[7] Those which have two syllables are often elided or rhyming, like iron and hi-fi. A few, like hoover, are named after the person or company who made the first successful version.

As yet, we have no short name for our digital assistants. 'Digital personal assistant' and 'virtual personal assistant' both capture the meaning but are hopelessly unwieldy. Maybe we'll initialise them, like TVs, and call them DAs, DPAs or VPAs. Or maybe we'll use the brand name of one of the early leaders, and call them all Siris. Google's chairman Eric Schmidt came up with the interesting idea that we'll find ways to name them after ourselves, and his would be called 'not-Eric'.[8] Perhaps we'll channel Philip Pullman's *His Dark Materials* trilogy and call them daemons, or perhaps – and this is my favourite – we'll just call them our 'Friends'.

4.3.4 Not by Voice Alone

We will probably also need to invent a new type of interface to enable us to communicate with our digital assistants. The 2013 movie *Her* is one of Hollywood's most intelligent treatments of advanced artificial intelligence (AI). (I realise that isn't saying much, but for better or worse, Hollywood does give us many of the metaphors that we use to think about and discuss future technologies.) The essence of the plot is that the hero falls in love with his digital assistant, with intriguing consequences. Although he uses keyboards occasionally, most of the time they communicate verbally.

There will be times when we want to communicate with our 'Friends' without making a sound. Portable 'qwerty' keyboards will not suffice, and virtual hologram keyboards may take too long to arrive – and they may feel too weird to use even if and when they do arrive. Communication via brain–computer interfaces will take still longer to become feasible, so perhaps we will all have to learn a new interface – maybe a one-handed device looking something like an ocarina.[9]

Another possibility is sensors – perhaps embedded as tattoos – on the face and around the throat which have microsensors to detect and interpret the tiny movements when people sub-vocalise, that is, speak without actually making a noise. Motorola, now owned by Google, has applied for a patent on this idea.[10] Perhaps when we are out and about in the future, we will get used to seeing other people silently miming speech as we all chat happily to our 'Friends'.

Another way we may communicate with our Friends, and indeed with many of the newly intelligible objects in the IoT, is radar. In May 2015, Google posted a video to introduce Soli, a project which embeds a sophisticated radar sensor in a tiny chip. It uses no lenses, and there is nothing to break. It generates a virtual tool in the space above or in front of itself, a way to interpret human intent by tracking the tiniest motion of the human hand and fingers. Soli generates virtual representations of controls we are all familiar with, such as volume knobs, on–off buttons and sliders.[11]

4.3.5 Doing Business with Friends

Digital assistants will be very big business, and the evolution of their industry will be fascinating. Will it turn out to be a natural monopoly, where the winner takes all? If so, the winner will find itself the subject of intense regulatory scrutiny, and probably of moves to break it up or take it into public ownership. Or will there be a small number of immensely powerful contenders, as in the smartphone platform business, where Apple and Android have the field almost to themselves?

Will we all choose one brand of 'Friend' at an early age, or during adolescence, and stick with it for life, as many people do with smartphones? Doubtless, the platform providers will seek to lock us in to that kind of loyal behaviour. Or will we be promiscuous, hopping from one provider to the next as they jostle and elbow each other, taking turns to launch the latest, most sophisticated software?

4.3.6 Wearables, Insideables

At the moment, the vessel which transports the primitive forebears of these essential guides is the smartphone, but that is merely a temporary embodiment. We will surely progress from portables to wearables (Apple Watch, Google Glass, smart contact lenses ...) and eventually to 'insideables': sophisticated chips that we carry around inside our bodies.

You doubt that Google Glass will make a comeback? The value of a head-up display, where the information you want is displayed in your normal field of vision, is enormous; that's why the US military is happy to pay half a million dollars for each head-up display helmet used by its fighter aircraft pilots.

Apple Watch has been successful because some people will pay good money to simply raise their wrist rather than go to all the bother of pulling their smartphone out of their pocket. How much better to have that hunger for the latest bit of gossip sated, and that essential flow of information about your environment displayed right in front of your eyes with no effort whatsoever?

Maybe the company which produces the first successful smart glasses for consumers will be Amazon. Its attempts to launch a smartphone failed, but its Echo product range succeeded spectacularly in creating the market for voice-activated smart speakers for the home, powered by Alexa, its digital assistant. Significantly, Amazon hired Babak Parviz, founder of Google Glass, in 2014, along with several other glass researchers, engineers and designers. Rumours suggest that Amazon might initially release smart glasses without the cameras and screens that inspired such privacy concerns, and rely solely on voice input.[12]

With regard to 'insideables', the technology to enable a chip implanted inside you to project imagery into your field of vision is far ahead of where we are now. But it is the next logical step in the process after wearables, and with key aspects of technology advancing at an exponential rate it would be foolish to write it off.

Screens will be everywhere by this time, of course: on tables, walls both interior and exterior, on the backs of lorries so that you can see what is ahead of them.[13] But we will probably want to carry our own screens around with us, not least because we won't always want other people to see what we are looking at.

In the coming decade, AI will have huge impacts on virtually every industry. By way of example, let's look at one industry which has so far lagged behind the general trend to improved performance from better information: healthcare.

4.3.7 Healthcare

It has been observed that our healthcare systems are really sick-care systems, often spending 90% of the amount they ever spend on an individual during the final year of their lives. We all know that prevention is better than cure, and that problems are most easily solved when identified early on, but we don't run our healthcare systems that way.

Two major revolutions are about to sweep across the healthcare horizon, and we will all benefit. One is the availability of small instruments which attach to our smartphones, enabling each of us to diagnose early symptoms of disease, and transmit relevant data to remote clinicians. These instruments are the result of cheaper and better sensors, and the application of AI algorithms and human ingenuity to huge data sets. They will cut out millions of time-consuming and expensive visits to doctors, and enable us to tilt sick-care towards healthcare.

The other revolution is the ability to anticipate and forestall medical problems by analysing our genomes. The Human Genome Project was completed back in 2003, but it soon turned out that although sequencing our DNA was an essential first step to enabling the practical improvements to healthcare we hoped for, it was not enough. We needed to understand epigenetics too: the changes in our cells that are caused by factors above and beyond our DNA sequence. The application of AI algorithms to the data which scientists are generating about gene expression are now bringing those improvements within reach.

There is almost no aspect of life today which is not being improved by AI. It is important to bear that in mind as we look at the potential downsides of this enormously powerful technology, and avoid a backlash which could prevent us benefiting from those improvements.

4.4 THE DISCOMFORT OF DISRUPTION

The exponential improvement of computing and AI will bring enormous benefits, but this means change, and change is uncomfortable.

4.4.1 Business Buzzwords

In the early 2010s, the hottest buzzword (buzz-phrase?) in business circles was Big Data. As always, there was a good reason for this, even if the endless repetition of the phrase became tiresome. Executives woke up to the fact that the huge amount of information they had about their customers and their targets could finally be analysed and turned into useful insights thanks to the reduced cost of very powerful (parallel processing) computers and clever AI algorithms. At long last, consumer goods manufacturers could do better than simply grouping their customers into zip codes, or broad socio-demographic tribes. They could find tiny sub-clusters of people with identical requirements for product variants and delivery routes, and communicate with them in a timely and accurate fashion.

4.4.2 Previous Disruption by Technological Innovation

In the mid-2010s, the buzzword became digital disruption, and again there is good reason for it. But the disruption of businesses and whole industries by technological innovation is nothing new.

On 13th October 1908, a German chemist named Fritz Haber filed a patent for ammonia, having managed to solidify nitrogen in a useful and stable form for the first time: three atoms of hydrogen and one of nitrogen. Nitrogen is a basic nutrient for plants, and it makes up 78% of the atmosphere, but its gaseous form is hard for farmers to use.

While Haber was developing his technique, thousands of men were labouring in the hot sun of Chile's Atacama Desert, digging saltpetre (nitrate) out of the ground. Apart from guano from Peru, Chile's saltpetre was the world's only source of nitrogen in solid form. As soon as Haber announced his discovery, saltpetre became uneconomic, and if you visit the Atacama today you can still see trains that were carrying loads of it to the port which were simply abandoned mid-journey, and substantial ghost towns which were abandoned in October 1908.

Today's disruption is caused by the digital revolution – the Internet, to be specific – and it is unusual in that it is affecting so many industries simultaneously. During the dot-com boom and bust at the turn of the century, there was much talk of companies being 'disintermediated' by the Internet, and a continuing example of that is the publishing industry, where publishers no longer dictate whose books get read because Amazon has enabled authors to publish themselves.

4.4.3 Kodak

The poster child of digital disruption is Kodak. At its peak in the late 1980s, it employed 145,000 people and had annual sales of $19 billion. Its 1,300-acre campus in Rochester, New York had 200 buildings, and George Eastman was as revered there a few decades ago as Steve Jobs is in Silicon Valley today.

Kodak's researchers invented digital photography, but its executives could not see a way to make the technology commercially viable without cannibalising their immensely lucrative consumer film business. So other companies stepped in to market digital cameras: film sales started to fall 20%–30% a year in the early 2000s, even before the arrival of smartphones. Kodak is often accused of being complacent but the dilemma it faced was almost impossible. In the classic phrase of the dot-com era, it needed to eat its own babies, and this it could not do.

Instead, it spent a fortune entering the pharmaceuticals industry, paying $5.1 billion for Sterling Drug, and another fortune entering the home printing industry. Both ventures were unsuccessful, and Kodak filed for bankruptcy in 2012, emerging from it a year later as a mere shadow of its former glory. Today it has annual sales of $2 billion and employs 8,000 people. Eighty of the 200 campus buildings have been demolished and 59 others have been sold off. Its market capitalisation at the time of writing is about $278 million, one-fifth that of GoPro, a maker of (digital) cameras for extreme sports that was founded in 2002.

4.4.4 Peer-to-Peer

A new business model which is generating a lot of column inches for the idea of digital disruption is peer-to-peer commerce, the leading practitioners of which are Airbnb and Uber. Both were founded in San Francisco, of course – in 2008 and 2009 respectively.

The level of investor enthusiasm for the peer-to-peer model is demonstrated by comparing Airbnb's valuation of $31 billion in March 2017 with Hyatt's market cap of $7.7 billion. Hyatt has over 500 hotels around the world and revenues of $4.4 billion. Airbnb, with 13 members of staff, owns no hotels and its revenues in 2016 were around $1.7 million. Uber's rise has been even more dramatic: its valuation reached $68 billion in June 2017, although this valuation was undermined by a string of negative stories.

This sort of growth is unsettling for competitors. Taxi drivers around the world protest that Uber is putting them out of business by competing unfairly, since (they claim) its drivers can flout safety regulations with impunity. Hoteliers have tried to have Airbnb banned from the cities where they operate, sometimes successfully. The growth of the peer-to-peer giants has not been smooth, of course. Uber especially has been embroiled in numerous spats with local authorities in the territories where it wanted to operate, and accusations of corporate sexism became so distracting that its founder and CEO was forced to resign.

A sub-industry of authors and consultants has sprung up, offering to help businesses cope with this disruption. One of its leading figures is Peter Diamandis, who is also a co-founder of Silicon Valley's Singularity University. Diamandis talks about the Six Ds of digital disruption, arguing that the insurgent companies are

1. Digitised, exploiting the ability to share information at the speed of light.

2. Deceptive, because their growth, being exponential, is hidden for some time and then seems to accelerate almost out of control.

3. Disruptive, because they steal huge chunks of market share from incumbents.

4. Dematerialised, in that much of their value lies in the information they provide rather than anything physical, which means their distribution costs can be minimal or zero.

5. Demonetised, in that they can provide for nothing things which customers previously had to pay for dearly.

6. Democratised, in that they make products and services which were previously the preserve of the rich (like cell phones) available to the many.

The task for business leaders is to work out whether their industry can be disrupted by this sort of insurgent (hint: almost certainly yes) and whether they can do the disruption themselves rather than being left standing in rubble like Kodak.

Digital disruption isn't devastating only because it enables competitors to undercut your product and service price so dramatically. That cheapness also means there will be many more potential disrupters because the barriers to entry are disappearing. Small wonder that Monitor, the business consultancy established by Michael Porter to advise companies how to erect those barriers, went bankrupt.

Business leaders often know what they need to do: set up small internal teams of their most talented people to brainstorm potential disruptions and then go ahead and do the disrupting first. These teams need high-level support and freedom from the usual metrics of return on investment, at least for a while. The theory is fairly easy but putting it into practice is hard: most will need external help, and many will fail.

Of course, the disrupters can also be disrupted. A service called LaZooz,[14] based on blockchain technology, may provide serious competition for Uber.

As computing and AI improve exponentially, the excitement and discomfort of disruption is only going to increase.

4.5 RELATED TECHNOLOGIES

4.5.1 One Ring to Bind them

AI is increasingly our most powerful technology, and it will increasingly inform and shape everything we do. Its full-blooded arrival coincides with the take-off of a series of other technologies. They are often driven at least in part by AI, and they will all impact the way our societies evolve.

Because they will all unfold in different ways and at different speeds, it is impossible to predict exactly what the individual and combined impact of these interlacing technologies will be, other than that it will be profound.

4.5.2 The Internet of Things

IoT has been talked about for years – the term was coined by British entrepreneur Kevin Ashby back in 1999.[15] Indeed, it has been around for long enough to have acquired a selection of synonyms. GE calls it the 'industrial Internet', Cisco calls it the 'Internet of everything', and IBM calls it 'smarter planet'. The German government calls it 'industrie 4.0',[16] the other three being the introduction of steam, electricity and digital technology. (I will explain in Section III of this book that I think this is an unhelpful term, as it shifts the IoT from the information revolution to the industrial one, and it understates the importance of the information revolution.)

My favourite alternative name for the IoT is ambient intelligence,[17] which comes nearest to capturing the essence of the idea, which is that so many sensors, chips and transmitters are embedded in objects around us that our environment becomes intelligent – or at least, intelligible.

When originally conceived, the IoT was based on radio frequency identification (RFID) tags, tiny devices about the size of a grain of rice which can be 'read' remotely without being visible to the device which 'reads' them. The RFID is a passive device, and this concept does not involve any AI.

Later, technologies like near field communication (NFC) were developed, which allow for two-way data exchange. Android phones have been NFC-enabled since 2011, and it powers the Apple Pay system which was launched with the iPhone 6.

The IoT is becoming possible because the component parts (sensors, chips, transmitters, batteries) are becoming cheaper and smaller at – yes – an exponential rate. A report published in January 2017 forecast a total of 23 billion IoT devices in 2021, up nearly four times on the 2016 figure.[18] Many of these devices have multiple sensors – smartphones can have as many as 30 each.[19]

Looking further ahead, the Internet entrepreneur Marc Andreessen predicts that by 2035, every physical item will have a chip implanted in it. 'The end state is fairly obvious – every light, every doorknob will be connected to the Internet'.[20]

Making the environment intelligible offers tremendous opportunities. A bridge, building, plane, car or refrigerator with embedded sensors can let you know when a key component is about to fail, enabling it to be replaced safely without the loss of convenience, money or life which unforeseen failure might have caused. This is known as condition-based maintenance, or predictive maintenance, and is being pioneered with encouraging results by, for instance, MTR Corporation, which runs Hong Kong's urban transit network.[21]

The IoT will also improve energy efficiency across the economy, as the heating or cooling of buildings and vehicles can be regulated according to their precise temperature, humidity and so on, and the number and needs of the people and equipment using them.

Since its launch in 1990,[22] the World Wide Web has rendered our lives immeasurably easier, by placing information at our fingertips. The IoT will take that process an important stage further, by dramatically improving the amount and quality of information, and enabling us to control many aspects of our environment. You will be able to find out instantaneously the location and price of any item you want to buy. You will know the whereabouts and welfare of all your friends and family – assuming they don't mind – and the location of all your property: no more lost keys! You will be able to control at a distance the temperature, the volume, the location of things that you own. Your own health indicators can be made available to anyone you choose, which will certainly save many lives.

Many of the applications made possible by the IoT will be surprising. For instance, perhaps it will transform our approach to the punishment of crime. It is obvious to most people that our current approach is not working. One in every 1,000 adults in the United States is currently in jail, and recidivism rates show that all this achieves is to condemn many of them to a life (often short) of crime and drug addiction. Supporters of the status quo argue that the system keeps the public safe by removing criminals from general circulation, and it provides a measure of justice, but few would argue that it is good at reforming behaviour or rehabilitation. Reformers argue that removing the liberty of criminals is sufficient punishment, and it would be better to do that without also institutionalising them and driving them back into crime when released. They believe that restorative justice, in which criminals do as much as possible to undo the harm they have done, is better than purely retributive justice.

IoT technologies like wearable devices which cannot be removed, coupled with sophisticated sensors in the immediate environment, could be deployed to restrict the freedom of convicted criminals, while allowing them to live in humane conditions which would make it less likely that they will reoffend after their term is complete. The idea is controversial, with some complaining that prisoners will have 'cushy' sentences, and others fearing the arrival of Big Brother. But the technology at least allows discussion of alternatives to a system that is currently broken.[23]

Like any powerful technology, the IoT will raise concerns, particularly about privacy and security, and we will return to these later. It will also need a set of standards, so that all those semi-intelligent chairs and cars talk the same language. This may come about through government regulation, industry cooperation, or because one player becomes strong enough to impose its standards on everyone else.

4.5.3 Robots

The final round of the Defence Advanced Research Projects Agency (DARPA) Robotics Challenge in June 2015 could have been a triumphal display of engineering prowess and the potency of AI. (For what are robots, but the peripherals of AI systems, just as mouses and keyboards are the peripherals of PCs?) Instead, as we noted previously, it was a sad affair, with the winning machine taking nearly 45 minutes to complete a series of eight tasks that a toddler could accomplish in 10 minutes. Moreover, the tasks had already been scaled down from the initial targets set in 2012 when it became obvious that none of the teams were going to be able to achieve them.[24]

But remember the progress in self-driving cars provoked by the DARPA Grand Challenge. In the initial event in 2004, the winning car drove just seven of the 150 miles of the track before crashing. A mere 13 years later, self-driving cars are demonstrably superior to human drivers in almost all circumstances, and they are closing the remaining gap fast. As far as robotics is concerned, we are at 2004 again. And don't forget the power of exponential improvement.

In Chapter 12: The History of Automation, we will meet Baxter, a new generation of industrial robot, which is beginning to demonstrate that robots can be flexible, adaptable and easy to instruct in new tasks. Research teams around the world are teaching robots to do intricate tasks. In October 2015, a consortium of Japanese companies unveiled the Laundroid, a robot capable of folding a shirt in four minutes.[25] Meanwhile at the University of California, a team developing the Berkeley robot for the elimination of tedious tasks (BRETT) spent 7 years reducing the time to fold a towel from 20 to 1.5 minutes.[26]

So, robots can fold towels – slowly, but it will be a few years before they can carry out efficiently all the tasks that a hotel chambermaid does. What they can already do, however, is automate many of the individual tasks that chambermaids carry out. Half a dozen big-name hotels in California are experimenting with robots that deliver towels and other items to guests' rooms on demand.[27] Apparently, toothpaste is the most-often requested item, presumably to keep all those perfect Hollywood teeth sparkling.

In mid-2015, a team at University of California, Berkeley, announced that by applying deep learning to the problem, they could get robots to screw tops onto bottles and remove nails from wood with a claw hammer and do so with approximately the same speed and dexterity of a human.[28]

Researchers are trying out different ways to improve robot performance. Teams at Carnegie Mellon University in Pittsburgh and at Google are getting robots to learn about their physical environment by having them simply prod, poke, grasp and push objects around on a tabletop, in much the same way that a human child learns about the physical world. Having collected a large data set from this activity, the systems turn out to be better at recognising images from the ImageNet database than systems which have not had the physical training.[29]

4.5.4 Google's Robot Army Decamps to Softbank

In late 2013, Google announced the purchase of no fewer than eight robotics companies. (Since you ask, they are Boston Dynamics – purveyor of the famous Big Dog and Atlas models – Bot and Dolly, Meka, Holomni, SCHAFT, Redwood, Industrial Perception and Autofuss.) Google also announced that the new division which owned them would be run by Andy Rubin, who had created a huge global business with the Android phone platform.

A year later, in October 2014, Andy Rubin left Google to found a technology start-up incubator, and it became apparent that Google had not managed to integrate its robotic acquisitions satisfactorily. In particular, the military focus of Boston Dynamics sat uneasily with Google's culture. In June 2017, it was announced that the Japanese investment group Softbank had acquired both Boston Dynamics and Schaft. A Japanese technology firm, Softbank is Japan's third-largest public company. It had already acquired another leading robotics company, France's Aldebaran, maker of the cute-looking and child-sized Pepper robots, in 2012.[30]

Softbank's founder Masayoshi Son, believes that a singularity will happen by 2047. In January 2017, he announced the formation of Vision, an investment fund with $100 billion to invest in high-technology businesses in preparation for this event.[31]

4.5.5 Complicated Relationships

It is going to take us humans a while to get used to having robots around. A French company called Aldebaran, which is owned by Japanese firm Softbank, manufactures a robot called Pepper. One hundred and twenty centimetres tall and costing around $1,200, they have a limited ability to 'read' human emotions and respond appropriately. They have proved extremely popular in Japan, with four batches of 1,000 selling out in less than a minute when they went on sale in September 2015.

The response to Pepper has not been straightforward, however. The manufacturer felt obliged to outlaw any attempt to engage in sex with the robot, and a Japanese man was prosecuted for assaulting one when drunk.[32]

A robot called Hitchbot managed to cross Canada from coast to coast in 2015, but was attacked and decapitated in Philadelphia when it tried to repeat the performance in the United States.[33]

4.5.6 More Robots: Androids, Drones and Exoskeletons

It is not clear that robots need to resemble humans closely to perform their tasks, but that doesn't stop researchers from trying to make them. (Robots with human appearance used to be what the word 'android' meant before Google appropriated it for phone software.) We are probably quite a few years away from having robots with the verisimilitude of the ones in the film *Ex Machina*, or the TV series *Humans*, for example. Nadine is a state-of-the-art prototype working as a receptionist at Nanyang Technological University in Singapore. It is humanoid, but doesn't fool anyone who takes a second glance.[34] Modelled on its inventor, Professor Nadine Thalmann, it cannot walk, but it can smile, turn its head and shake your hand. Its voice is powered by an AI similar to Siri.

Many robots will be special-purpose devices, constructed to carry out a very specific task. An example is the Grillbot, a robot the size of a table tennis bat which cleans your barbecue grill, and is otherwise entirely useless.[35]

Some people argue that exoskeletons are wearable robots. Whether or not that is semantically correct, they will certainly enable one human to do the work of several. At the moment, leading companies in the space like Ekso Bionics[36] are focusing on patient rehabilitation systems. But before long similar equipment will be available for people carrying out physically demanding tasks in the military, manufacturing and distribution.

4.5.7 Drones

Another form of robot which is taking off fast is drones – flying machines that can be controlled remotely or autonomously. They have a wide range of applications, including taking surreptitious photos of celebrities, taking selfies for life-logging millennials, and delivering parcels for Amazon. They present a serious challenge for regulators concerned about the impact on more established forms of aircraft, but they cannot be wished away: Internet-connected drones with powerful sensors and computers on board are quickly becoming essential tools for companies in the utilities and engineering industries, as well as government agencies.[37] In Switzerland, a drone company called Matternet received approval in 2017 to start flying medical supplies to hospitals in urban areas aboard robotic quadcopters.[38] And the Icelandic Transport Authority gave a company called Flytrex approval to fly drones along an out of line-of-sight route. Initially, customers would have to pick up deliveries from a designated drop-off zone, but the company expected approval shortly to fly direct to their backyards.[39]

4.5.8 Virtual Reality

During 2014, many people got their first taste of VR from Google Cardboard, an ingenious way to let smartphones introduce us to this extraordinary technology. The year 2016 was widely expected to be the year that VR really took off, as Facebook's Oculus VR launched Rift, the first VR equipment for consumers that offered high-definition visuals and no latency. Latency is a failure of synchronisation between the stimuli from different sources reaching the brain: if your visual experience is out of synch with your other senses, your brain gets confused and unhappy, and can make you feel surprisingly sick.[40]

When VR is effective, it is surprisingly powerful. When the sense data being received by the brain become sufficiently realistic, the brain 'flips', and decides that the illusion being presented is the reality.

In the event, Oculus Rift was not the breakthrough product that many expected. Consumers felt the equipment was expensive and heavy to wear. VR rivals Sony and HTC fared better, but overall, the installed base remained small in early 2017, at less than 0.5% of the number of accounts on Steam, the biggest games distribution platform.[41]

Augmented reality (AR) is similar to VR except that it is overlaid on your perception of the real world rather than replacing it. It can make elephants swim through the air in front of you, or plant a skyscraper in your back garden. This is handy if you want to remain alert to the threat from dogs and potholes while you are hallucinating swimming elephants.

In 2016, AR stole the limelight from VR when a developer called Niantic achieved startling success with an AR version of Pokémon Go. A week after its launch in July 2016, 28 million people were using it every day, looking for virtual Pokémon characters to capture. The app was downloaded 650 million times by February 2017,[42] but the enthusiasm could not be sustained, and by September it was not among the iPhone's top 200 apps.[43]

Microsoft attempted to mark out a distinct territory for its own AR offering, the Hololens, but called it mixed reality. In mixed reality (which is not usually abbreviated to MR), the virtual object is anchored in the real world so that you can, for instance, walk around it and manipulate it. In many forms of AR, you only see the virtual object when you look at the screen of your smartphone. Hololens requires a headset, so the object is always present in your field of view. A start-up called Magic Leap raised $1.3 billion from investors including Google, with a series of impressively realistic demos.[44]

VR and AR enthusiasts are long-suffering and persistently optimistic. Excitement at the time of writing centres on Apple, which launched AR development software called ARKit in June 2017. The software that enables consumers to use the AR was unveiled in a surprisingly downbeat fashion in September 2017, perhaps to avoid the backlash of disappointment which has hit previous incarnations. But with an installed base of millions of potential users – far more than is currently available for Google's Tango, Apple's ARKit may still turn out to be the real breakthrough.

Insofar as there is any debate about whether VR is going to be an important development, it's between those who think it's going to be huge, and those who think it's going to change everything. Digi-Capital, a specialist consultancy, expects VR and AR sales to exceed $100 billion by 2021, with AR contributing four-fifths of the total.[45]

4.5.9 Applications of VR and AR

The biggest application in the short term is expected to be video games, which is no small playing field, since gaming has for some time rivalled Hollywood for leadership in global sales of packaged entertainment.[46] Judging by the content already being made available for Google Cardboard, people also enjoy ersatz travel and adventurous experiences. VR versions of Google Street View let you wander around Manhattan until the latency makes you ill, and other developers offer you rollercoaster rides, and adventure sports from skiing to hang gliding.

In the longer term, the potential applications are bewildering. Without ever leaving our armchairs we may soon be able to enjoy such realistic simulations of events like sports matches and music concerts that many people will question the value of struggling with transport and crowds to attend the real thing. Of course, the crowd has a lot to do with making the event exciting in the first place, so the organisers of VR events will want to find a way to recreate the effect of being in a crowd. Except that you'll be able to sit next to your friend, who happens to be in a VR rig a couple of continents away at the time.

Education and informal learning is also likely to experience a VR revolution. How much more compelling would it be to learn about Napoleon by experiencing the battle of Waterloo than by reading about it, or listening to a lecturer describe it? How much easier would it be for a teacher to explain the molecular structure of alcohol by escorting her pupils round a VR model of it?

Businesses will find many uses for VR, and because they often have larger budgets than consumers and educational institutions, they may sponsor the creation of the most cutting-edge applications. Computer-aided design environments will become startling places to work, for instance, allowing architects, designers and clients to explore and discuss buildings in great detail before ground is broken. And who knows what uses the military will find for VR. One frightening thought is that VR could become a powerful and truly terrifying instrument of torture.[47]

Telecommunication will also be taken to a new level. Although audio-only phone calls still predominate, good video-conferencing facilities add enormously to the effectiveness of a long-distance conversation, and the additional step of feeling present in the same space will improve the experience again. Anything which involves your relocation in time or space should be fertile ground for VR.

On the other hand, it is not yet clear whether VR will turn out to be a good medium for movies. In a film, the director wants to direct your attention, and it isn't helpful if half the audience is busy gawping at images or events 180 degrees removed from the focus of the action.

Cynics will point out that new media (TV, video, the Internet were all new media in their early days) are established only when users have found ways to apply them to porn, gambling and then sport. No doubt VR will make its contribution to these areas of human activity, but I'm not going to get sucked into a discussion of what could be achieved with haptic suits – clothing which allows users to experience sensations of heat and touch initiated remotely by someone else.

The death of geography has been declared numerous times, but despite the rise of telephony, digitisation and globalisation, business and leisure travel just keeps on growing. Could the sense of genuine 'presence' which good VR confers finally make the old chestnut come true? Will talent continue to be drawn into the world's major cities, or will VR puncture their inflated real estate prices, and smear humanity more evenly across the planet?

Maybe VR can render scarcity less valuable, and less problematic. In the real world, not everyone can live in a beautiful house on a palm-fringed beach, drive an Aston Martin, and be greeted by a Vermeer as they enter their living room. With VR, everyone can – to a fair degree of verisimilitude. As we will see in Chapter 14: The Challenges, this might turn out to be extremely important for our overall well-being as a species.

The World Wide Web has given us something like omniscience, and VR looks set to give us something like omnipresence. Perhaps all we need now is a technology to give us something like omnipotence.

4.6 CURRENT CONCERNS

Powerful new technologies can produce great benefits, but they can often produce great harm. Long before we get to the really big challenges posed by the two singularities, people have raised a series of significant concerns about AI, including privacy, transparency, security, bias, inequality, isolation, oligopoly, killer robots and algocracy. In no particular order, let's look at each of them.

4.6.1 Privacy

AI systems have huge appetites for data, and the IoT will generate it by the bathful. In an intelligent environment, the past and present location of every citizen is easy to establish, along with who they have met and very possibly what they discussed. Many people are understandably concerned about this information being used and misused by all sorts of organisations, including (depending on their political persuasion) governments, corporations, pressure groups and resourceful individuals – such as jealous spouses. As one group of activists puts it, the net is closing around us and we are increasingly transparent to organisations which are increasingly opaque to us.[48]

Some people hope that we can retaliate against this kind of 'surveillance on steroids' with 'sousveillance'. With cameras ubiquitous – including on drones – the actions of those in authority are constrained because they know that their actions are observed and recorded by members of the public. This is already happening with law enforcement, with police officers in the United States being prosecuted for harassment in situations where they would previously have been immune from oversight. Some authorities are actively embracing this development, with police officers being required to wear cameras at all times in order to pre-empt false allegations. With cameras on drones, the reach of civilian oversight can be extended so far that some are calling it 'Little Brother'.[49] With the watchers being watched, we may arrive at a balance called 'co-veillance'.[50]

The arms race over data will continue between governments, large organisations and the rest of us. Hollywood loves the trope of the socially dysfunctional hacker who is smarter, more up-to-date and more motivated than her opposite numbers in the civil service, but perhaps we should not be comforted by that idea. When forced to choose between privacy and the opportunity to share, we generally choose to share. We leave a trail of digital breadcrumbs wherever we go, both in the real world and online, and most of us are careless about it.

In part, this is because many of us feel that we have nothing to worry about because we have nothing to hide. But there is a chilling effect on free speech if we start to censor ourselves because we want to stay that way. We might think twice before entering a certain term into a search engine, or hesitate before making friends with someone who is overtly counter-cultural. Recent research shows that we self-censor when we are aware of the possibility that we are being surveilled, even when we know we are saying and writing nothing illegal.[51]

In 2015, the Chinese government provided an unnerving demonstration of where this could lead. It is building a 'social credit' database of all citizens which ranks them according to their trustworthiness. In a frightening extension of credit scoring systems, the database will incorporate all the financial and behavioural information the government can accumulate, and distil it into a single number, ranging from 350 to 950. A score above 600 qualifies a citizen for an instant loan worth $800. At 650 you can rent a car without leaving a deposit. At 700 you are fast-tracked for a Singapore travel permit. Important jobs will require high scores.

Citizens will earn demerits for reprehensible shopping habits (too many video games? too much wine?) and merits for socially responsible actions, like reporting bad behaviour by others.[52] A particularly scary aspect of the system is that people receive demerits if their friends on social media are marked down. The system will be compulsory for every Chinese citizen in 2020, and until then, eight pilots are being run by Chinese companies, including Sesame Credit, the financial wing of Alibaba, which is China's version of Amazon.

The US civil liberties pressure group ACLU thinks 'China's nightmarish Citizen Scores are a warning for Americans. ... There are consistent gravitational pulls towards this kind of behaviour on the part of many public and private US bureaucracies, and a very real danger that many of the dynamics we see in the Chinese system will emerge here over time'.[53]

Big data and AI could enable governments to build apparatus of control which would make Big Brother in George Orwell's '1984' look amateurish.

You're not necessarily safe from this prospect just because you live in a multiparty democracy. In April 2014, Nicole McCullough and Julia Cordray founded Peeple, an app which will enable people to rate each other according to their courtesy and helpfulness. Originally conceived as a way to improve behaviours, it was widely criticised as likely to become a medium for personal attack and bullying. The founders responded by changing the rules so that subjects would have a veto over any comments made about them on the site, although they left open the possibility that users who pay extra could see uncensored inputs. The watered-down version of the app launched in March 2016, but made little impact. At the time of writing, its rating on iTunes is 1.5 out of five.[54]

Clearly, we still have much to learn about how to conduct ourselves individually and collectively in the new world of data tsunamis and massive analytic horsepower.

Beyond the fact that we have fallen into careless habits with regard to our personal data, and we keep voting with our keystrokes to surrender ever more of it, there may be very good societal reasons to think that privacy cannot be retained, or rather restored. Technology is usually a two-edged sword, and the ability to manufacture deadly pathogens, deploy fearsome miniaturised killer drones and unleash devastating digital viruses is becoming ever cheaper. What price privacy when the cost of a megadeath falls to within the budget of a gruntled teenager? Perhaps we will have no sensible option but to surrender our privacy, and rather than keeping secrets, our right would be to know what our secrets are being used for.[55] These are complex issues, and will be debated over and over as new possibilities and threats open up.

Researchers at Google and Microsoft are experimenting with promising approaches to squaring the circle of protecting privacy while sharing data. Working with Cornell University in New York, Google is trying to enable groups of organisations (e.g. hospitals) to train deep learning algorithms on their own separate data files and then share the outputs of the trained data. They have found that this can work almost as well as combining all the data into one file and using that to train the algorithm.

Microsoft is using a technique called homomorphic encryption to perform analysis on data which is encrypted. It yields encrypted results which can then be decrypted without the sensitive data ever having been available to the analysts in unencrypted form.[56]

4.6.2 Transparency

AI systems, and especially those using deep learning, are often described as 'black box' systems. They generate effective and helpful results, but we cannot see how they arrived at their answers. This is a problem when important decisions are being made about people's lives. 'The computer says "no"' is an unacceptable response when you ask why your loan application was refused, or why your application for early release from jail was denied.

The financial services industry and the military are two sectors where this problem is particularly acute, for obvious reasons, and work is under way there as well as elsewhere to try to make deep learning systems more transparent.[57] If it succeeds, a world where many decisions are taken by

AIs will be more transparent than the one we live in today, where bureau-cracies are often unwilling or unable to tell us why particular decisions have been made. Humans often make decisions without being able or will-ing to explain their real motivations in full.

4.6.3 Security

In his book 'Future Crimes', security expert Marc Goodman sets out in detail how criminals, governments, and organisations use the swelling oceans of data being transmitted about us to steal from us and manipu-late us. Cybercrime is probably the fastest-growing type of crime all over the world; much of it goes undetected, and much of what is detected goes unsolved.

Another growing concern about hacking is sabotage. As the IoT is built out and more and more of our vehicles, buildings and appliances rely on AI, the problems that can be caused if their control systems are hacked increase in significance. The possibility of a hacker gaining control of every self-driving car in a city and making them all turn left at the same moment is frightening.

Programmers say that there is no such thing as 100% security: IT sys-tems are designed by humans, and we are fallible. They are also increas-ingly opaque and hard to debug. An optimist would say that although complex, well-defended systems come under frequent attack, they are rarely successfully hacked. No hacker has yet launched a US nuclear missile, although of course that doesn't mean that it will never happen. Eternal vigilance is the price we must pay to avoid disaster, and we are not practising it at the moment. Many of us are lax about safeguarding our Internet passwords, and many companies' security arrangements also fall far short of best practice.

Policemen say that when they are pursuing a criminal, the criminal needs to be lucky all the time whereas the police only need to get lucky once. But when the criminals are on the offensive, looking for security gaps, the boot is on the other foot.

In September 2016, the world was shown how much of a security risk the IoT can be when a 'botnet' called Mirai appeared. By taking control of multiple IoT devices, it launched a 'denial of service' attack against Dyn, a company that provides much of the Internet's infrastructure.[58] (A botnet is a collection of devices infected with malicious software, or malware, and used without their owner's knowledge. A denial of service attack floods a target computer with so many requests that it cannot function normally, and a distributed denial of service attack launches the attack from multiple computers simultaneously.) Observers noted that this showed how the IoT could be weaponised, and at the time of writing, governments are still trying to work out how to respond.[59]

4.6.4 Bias

We tend to think of machines as cold, calculating and unemotional, which of course they are. But they are also prejudiced. Not because of any implicit beliefs of their own – they have none – but because they pick prejudices up from us. Machines learn about the meanings and associations of words from the sentences that we give them, and bias is ingrained in human thought and language. Machines build up mathematical representations of language, in which meaning is captured by numbers, or vectors, based on the other words most commonly associated with them. Sometimes this is innocuous, as when we associate flowers with positive feelings and insects and spiders with negative ones. But it can have unwanted implications when, for example, we associate 'male' with professor and 'female' with assistant professor.[60]

An extreme example of a biased machine was provided by Microsoft in March 2016 when it launched a chatbot called Tay on Twitter. Within 24 hours, mischievous (and perhaps in some cases vicious) humans had converted Tay into an aggressive racist, and Microsoft, horrified, shut it down.

Inadvertently infecting our machines with existing bias is not necessarily making our world worse: in many cases it is simply perpetuating the problem that already exists. And at least machines will not attempt to cover up or post-rationalise their bias. If we can observe it we can tackle it, and by tackling the bias in machines, perhaps we can surface and tackle the bias in ourselves.

4.6.5 Inequality

Every time a new technology is launched, people worry that only the rich will have access to it, and there will be a 'digital divide' separating the haves and the have-nots. The life experiences and opportunities of the wealthy will diverge unacceptably from those of the rest of us.

So far, while not groundless, this fear has been exaggerated. It is true that in recent years the super-rich have gained more income and wealth than anyone else in most developed economies. (And we're not talking about the 1% here, but the 0.01%.) Meanwhile, there are people who struggle to afford what many would consider the basic necessities of life – although the definition of basic necessity varies greatly between developed countries and elsewhere in the world.

It is also true that the disparity of income between average people in rich countries and average people in the poorest countries is enormous. This disparity, however, is shrinking. And those in America and Europe who protest about the obscene wealth of the 1% in their own countries seem curiously untroubled by the fact that they themselves are often among the richest 1% of the world's population.

Inequality can be overemphasised as a social good. There is a considerable amount of empirical evidence that we don't actually care about inequality as much as we like to think. What we really care about is fairness.[61] And it is often an odd sort of fairness at that. Professional footballers are paid daft amounts of money for possessing talents which are impressive, but don't seem to have much to do with the moral value of a human. Yet their good fortune provokes remarkably little protest. CEOs, on the other hand, are subjected to considerable abuse, even though if they do what they say they do, they create great wealth for shareholders, employees and tax authorities alike. The reason for the difference is probably that we don't mind stars being paid inordinate amounts if they really are stars – different from the rest of us. Many people are not convinced that CEOs are stars in that sense.

The new technologies which have emerged during the various stages of the industrial revolution have become available to most people in developed economies not long after they were invented. The car, the refrigerator, the washing machine, the TV, the home computer, the smartphone – all have gone through the same cycle. An expensive first version is launched which can be afforded only by the wealthy. It doesn't work very well, and is at least in part a status symbol. Very quickly, the technology improves and the price falls, and pretty soon the great majority of us have one.

The reason for this is simple economics. Companies can make far more money by selling lots of cheap smartphones (for instance) to everyone than by selling a few very expensive ones to the wealthy elite. And in a competitive economy, even if the first company to market is happy to make its money by scalping the rich, other companies will quickly come along to raise the quality and reduce the price. There is no 'fridge divide'; why should there be a 'digital divide'?

I said this fear was exaggerated 'so far'. In Section III of this volume we will see that there may be more grounds for concern in the not-too-distant future.

4.6.6 Isolation

Parents have long fretted about their teenage children spending long hours in antisocial isolation, hunched over a video game console. Wishing their kids would go outside and kick a ball around instead, they have agonised over a series of scares about the ill effects of video games, which allegedly make kids violent, stop them developing social skills, render them vulnerable to legions of grooming molesters, and give them impossibly short attention spans. And the blue light of the screen disrupts their sleep.

Meanwhile, the Flynn Effect describes the finding that IQ levels are increasing steadily each generation,[62] which should not be surprising when you consider the general trends towards less smoking, less drinking, better central heating, better food and better healthcare. And the fact that we are continually learning more about what works in education and what does not.

Humans are intensely social creatures. The need to belong to a tribe – to be accepted by it and perhaps to climb its hierarchy – is programmed deeply into us. Working together in tribes is how we survived in the savannah, surrounded by animals which were stronger and faster, with bigger teeth. Individuals who were cast out of their tribe quickly joined another one or – more likely – got eaten. It would be amazing if in a single bound, one generation of teenagers suddenly freed themselves from this evolutionary programming and isolated themselves in solitary pursuits.

And indeed, they haven't. The most popular video games are those which people can play together and incorporate into their social bonding activities. For teenagers, these activities are just as important as they ever were – and of course it is no less important that their parents be at least slightly appalled by them.

If and when the day comes when people can plug into utterly compelling VR worlds through a direct neural link, and effectively disappear into the Matrix, things may be different. But unless we have altered our cognitive make-up dramatically by then, my hunch is that we will find a way to make the Matrix social too.

4.6.7 Oligopoly: Regulators to the Rescue?

In the last few years, Silicon Valley engineers must feel as if they have gone from zeroes to heroes to villains. They may still feel that they are the lovable, downtrodden geeks from *the Big Bang Theory* TV series, but much of the media now paints them as 'brogrammers', a sort of tech version of frat boys: over-paid, arrogant and definitely too male.

In particular, many people are worried that the tech giants are too powerful and argue that they need to be weakened, restrained or controlled. At the time of writing, the combined market capitalisation of Apple, Google, Microsoft, Amazon and Facebook is about $3 trillion, which is greater than the market capitalisation of the FTSE-100, the index of the UK's 100 biggest companies. In technology, even more than elsewhere, success breeds success. In 2016, Apple earned over 100% of the profits generated in the smartphone industry. It only supplied 12% of the products sold, but everyone else lost money.[63] An analysis in October 2017 showed that the companies which led mobile computing from 2005 to 2016 (Google, Apple, Facebook and Amazon, or GAFA) had ten times the revenue of the companies (Microsoft and Intel, or Wintel) which led desktop computing from 1990 to 2001. The tech giants have stepped onto a bigger stage.[64]

Collectively, these six companies spend more on R&D than the UK government plus all the UK's companies and universities combined. This gives these companies great market power and the ability to influence political power.

The European Commission has been pondering this issue for years. It accuses the GAFA companies in particular of dodging taxes by booking revenues which are transacted in cyberspace to low-tax domiciles. In June 2017, the Commission fined Google €2.4 billion for favouring its own shopping comparison service in the results provided by its search engine. Although Google can easily afford a fine of this size, it was the largest ever imposed by the EU's competition regulators, and it was widely seen as a declaration of more general intent. As *the Guardian* newspaper put it, 'However benign their intentions, the sheer size and reach of these companies makes them dangerous. This judgement represents one of the few serious attempts to manage these monopolies. It's a welcome start'.[65]

Competition legislation is necessary, and monopolies should be broken up where they form and are shown to be operating against the public interest. But companies should not be penalised just because a lot of people think they are too big and too powerful. And the remedies should be fair and transparent. Google will almost certainly appeal against this ruling, and the debate about whether it is sensible, legal or just will probably rumble on through the courts for months or years. Perhaps, rather than drawing up battle plans which may be illegal and counterproductive, the Commission should take a leaf out of Denmark's book, and appoint an ambassador to GAFA.[66] As we will see in later chapters, the tech giants have an important role to play in getting us all through the two singularities, and governments should work with them to achieve that.

The tech giants employ clever lawyers, and take full advantage of the new circumstances of online business to minimise their tax bills. Multinational companies have always presented challenges to revenue officials: in vertically integrated industries like oil and pharmaceuticals, transfer pricing decisions between subsidiaries have long had major impacts on the income streams of governments. But it is governments which set the rules under which taxes are paid, and it is hardly reasonable to expect people or companies to pay for the mistakes those governments have made, or for their failure to keep up with new realities.

Markets often achieve what regulators cannot and should not try to. It is true that there are network effects in information-based industries which can favour the emergence of monopolies. But there is also fierce competition, and business models change quickly, creating losers out of winners and vice versa. The histories of IBM, Microsoft and Apple illustrate this clearly, and Google, Amazon and Facebook are not immune. There are fears that this time it is different, and (for instance) that the post-IPO struggles of Snapchat show that disruption is over, and Facebook is unassailable,[67] but this is extremely unlikely.

Technology companies are perennially vulnerable to attack from innovators and from changes in customer behaviour which may undermine their services. Google makes most of its money from selling ads alongside search results, and that is vulnerable if consumers use different search engines, and if the nature of search changes.

Google does not see the primary competition to its search engine as Bing, Microsoft's search engine, or DuckDuckGo, a search engine that promises not to retain your personal information. Its real competition in search is Amazon. In August 2017, the cosmetics giant L'Oréal announced it was moving some of its search budget from Google to Amazon, saying that 38% of consumer searches for cosmetics started on Amazon.[68]

Perhaps even more threatening to Google is the way that search will change over the coming years as we control our computers more and more by voice rather than with keyboards. When you are looking at a screen it is easy to serve you ads along with the specific information you requested. It is hard to see how this can be replicated when the machine gives you the requested information verbally.

Regulation may be required at times, but because it tends to tackle issues which have already faded, it can have modest or even negative impacts, and should therefore be embarked upon with great caution.

4.6.8 Killer Robots

Human Rights Watch and other organisations are concerned that in the coming years, fully autonomous weapons will be available to military forces with deep pockets.[69] They argue that lethal force should never be delegated to machines because they can never be morally responsible. Their stance has garnered a great deal of support.

In his book *Homo Deus*, Yuval Hahari sets the counterargument with his customary genteel brutality: 'Suppose two drones fight each other in the air. One drone cannot fire a shot without first receiving the go-ahead from a human operator in some bunker. The other drone is fully autonomous. Which do you think will prevail? … Even if you care more about justice than victory, you should probably opt to replace your soldiers and pilots with autonomous robots and drones. Human soldiers murder, rape and pillage, and even when they try to behave themselves, they all too often kill civilians by mistake'.

Ultimately, the decision whether to develop and deploy these weapons will be taken by governments and their military personnel. It is logically possible that all governments and all military commanders everywhere will refrain in perpetuity, but it is hard to believe that will happen in practice.

4.6.9 Algocracy

The final concern we will review here receives far less attention than the others at the moment, but could turn out to be the most significant and the most lasting of them.

Decisions about the allocation of resources are being made all the time in societies, on scales both large and small. Because markets are highly efficient systems for allocating resources in economies characterised by scarcity, capitalism has proved highly effective at raising the living standards of societies which have adopted it. Paraphrasing Churchill,[70] it is the worst possible economic system except for all the others.

Historically, markets have consisted of people. There may be lots of people on both sides of the transaction (flea markets are one example, eBay is another). Or there may be few buyers and many sellers (farmers selling to supermarket chains) or vice versa (supermarket chains selling to consumers). But typically, both buyers and sellers were humans. That is changing.

Algorithms now take many decisions which were formerly the responsibility of humans. They initiate and execute many of the trades on stock and commodity exchanges. They manage resources within organisations providing utilities like electricity, gas and water. They govern important parts of the supply chains which put food on supermarket shelves. This phenomenon will only increase.

As our machines get smarter, we will naturally delegate decisions to them which would seem surprising today. Imagine you walk into a bar and see two attractive people at the counter. Your eye is drawn to the blond but your digital assistant (located now in your glasses rather than your phone) notices that and whispers to you, 'hang on a minute: I've profiled them both, and the red-head is a much better match for you. You share a lot of interests. Anyway, the blond is married'.

In his 2006 book *Virtual Migration*, Indian-American academic A. Aneesh coined the term 'algocracy'.[71] The difficulty with it has been explored in detail by the philosopher John Danaher, who sets the problem up as follows. Legitimate governance requires transparent decision-making processes which allow for involvement by the people affected. Algorithms are often not transparent, and their decision-making processes do not admit human participation. Therefore, algorithmic decision-making should be resisted.[72]

Danaher thinks that algocracy poses a threat to democratic legitimacy, but does not think that it can be, or should be, resisted. He thinks there will be important costs to embracing algocracy and we need to decide whether we are comfortable with those costs.

Of course, many of the decisions being delegated to algorithms are ones we would not want returned to human hands – partly because the machines make the decisions so much better, and partly because the intellectual activity involved is deathly boring. It is not particularly ennobling to be responsible for the decision whether to switch a city's street lights on at 6.20 or 6.30 p.m., but the decision could have a significant impact. The additional energy cost may or may not be offset by the improvement in road safety, and determining that equation could involve collating and analysing millions of data points. Much better work for a machine than a human, surely.

Other applications make us much less sanguine. Take law enforcement: a company called Intrado provides an AI scoring system to the police in

Fresno, California. When an emergency call names a suspect, or a house, the police can 'score' the danger level of the person or the location and tailor their response accordingly.[73] Other forces use a 'predictive policing' system called PredPol which forecasts the locations within a city where crime is most likely to be carried out in the coming few hours.[74] Optimists would say this is an excellent way to deploy scarce resources. Pessimists would reply that Big Brother has arrived.

AI is already helping to administer justice after the event. In 2016, the San Francisco Superior Court began using an AI system called PSA to determine whether parole should be given to alleged offenders. They got the tool free from the John and Laura Arnold Foundation, a Texas-based charity focused on criminal justice reform. Academics studying this area have found it very hard to obtain information about how these systems work: they are often opaque by their nature, and they are also often subject to commercial confidentiality.[75]

There are many decisions which machines could make better than humans, but we might feel less comfortable having them do so. The allocation of new housing stock, the best date for an important election, the cost ceiling for a powerful new drug, for instance. Arguments about which decisions should be made by machines, and which should be reserved for humans are going to become increasingly commonly and increasingly vehement. Regardless whether they make better decisions than we do, not everyone is going to be content (to paraphrase Grace Jones) to be a slave to the algorithm.

Information is power. Machines may intrude on our freedom without actually making decisions. In September 2017, a research team from Stanford University was reported to have developed an AI system which could do considerably more than just recognise faces. It could tell whether their owners were straight or gay. The idea of a machine with 'gaydar' is startling; it becomes shocking when you consider the uses it might be put to – in countries where homosexuals are persecuted and even prosecuted, for instance.[76] The Stanford professor who led the research later said that the technology would probably soon be able to predict with reasonable accuracy a person's IQ, their political inclination or their predisposition towards criminality.

These concerns are not trivial, but none of them entails a singularity. Before we discuss the developments that do, we need to understand more about that term.

NOTES

1. *The Economist*, December 4, 2003.
2. Douglas Adams, *The Salmon of Doubt*.
3. http://www.wired.com/2014/10/future-of-artificial-intelligence/.
4. https://www.gsb.stanford.edu/insights/andrew-ng-why-ai-new-electricity.
5. http://www.geektime.com/2017/07/30/what-is-the-future-of-chatbot-development-and-artificial-intelligence/.
6. http://www.bbc.co.uk/news/technology-40739709.
7. Not an everyday object outside the United States, of course.
8. http://www.bloomberg.com/news/articles/2016-01-11/google-chairman-thinks-ai-can-help-solve-world-s-hard-problems-.
9. An ocarina is a wind instrument about the size of a fist. First introduced to Europeans by the Aztecs, it looks like a toy submarine.
10. https://www.extremetech.com/extreme/171992-motorola-patents-e-tattoo-that-can-read-your-thoughts-by-listening-to-unvocalized-words-in-your-throat.
11. https://www.youtube.com/watch?v=0QNiZfSsPc0.
12. https://techcrunch.com/2017/09/20/amazon-is-working-on-smart-glasses-to-house-alexa-ai-says-ft/?utm_medium=TCnewsletter.
13. This is actually a great idea, which is being trialled in Argentina at the time of writing: http://www.telegraph.co.uk/motoring/motoringvideo/11680348/Transparent-trucks-with-rear-mounted-Samsung-safety screens-set-to-save-overtaking-drivers.html. Of course, it may be less valuable when cars drive themselves and their human occupants don't look at the road.
14. http://lazooz.org/.
15. http://www.rfidjournal.com/articles/view?4986.
16. http://www.vdi-nachrichten.com/Technik-Gesellschaft/Industrie-40-Mit-Internet-Dinge-Weg-4-industriellen-Revolution.
17. Coined by another British entrepreneur, Simon Birrell: https://www.linkedin.com/in/simonbirrell.
18. http://uk.businessinsider.com/the-internet-of-things-2017-report-2017-1?r=DE&IR=T.
19. http://singularityhub.com/2016/02/09/when-the-world-is-wired-the-magic-of-the-internet-of-everything/.
20. http://www.telegraph.co.uk/technology/internet/12050185/Marc-Andreessen-In-20-years-every-physical-item-will-have-a-chip-implanted-in-it.html.
21. http://www.information-age.com/it-management/strategy-and-innovation/123460379/trains-brains-how-artificial-intelligence-transforming-railway-industry.
22. http://home.cern/topics/birth-web.
23. http://www.abc.net.au/news/2017-08-14/how-ai-could-put-an-end-to-prisons-as-we-know-them/8794910
24. http://www.popsci.com/darpa-robotics-challenge-was-bust-why-darpa-needs-try-again.

25. http://uk.businessinsider.com/laundroid-japanese-robot-folds-laundry-2015-10.
26. http://www.npr.org/sections/money/2015/05/19/407736307/robots-are-really-bad-at-folding-towels.
27. http://www.techinsider.io/savioke-robot-butler-in-united-states-hotels-2016-2.
28. http://www.kurzweilai.net/the-top-ai-breakthroughs-of-2015.
29. http://www.nextgov.com/emerging-tech/2016/05/robots-are-starting-learn-touch/128065/.
30. https://www.recode.net/2017/6/8/15766440/softbank-alphabet-google-robotics-boston-dynamics-schaft.
31. http://uk.businessinsider.com/softbank-ceo-masayoshi-son-thinks-singularity-will-occur-within-30-years-2017-2.
32. http://www.theguardian.com/world/2015/sep/28/no-sex-with-robots-says-japanese-android-firm-softbank.
33. https://www.theguardian.com/technology/2015/aug/03/hitchbot-hitchhiking-robot-destroyed-philadelphia.
34. http://www.telegraph.co.uk/news/science/science-news/12073587/Meet-Nadine-the-worlds-most-human-like-robot.html.
35. http://techcrunch.com/2016/01/07/the-grillbot-is-a-robot-that-cleans-your-grill/#.w9z87m:Hd0d.
36. http://intl.eksobionics.com/.
37. http://singularityhub.com/2016/02/29/drones-have-reached-a-tipping-point-heres-what-happens-next/.
38. https://www.technologyreview.com/the-download/608912/urban-drone-deliveries-are-finally-taking-flight/.
39. https://www.technologyreview.com/the-download/608718/finally-theres-a-halfway-compelling-consumer-drone-delivery-service/.
40. Your brain is wired to make you see things before you hear them as it knows that light travels faster than sound. Thus, the brain can tolerate audio lagging video, but is much less tolerant of video lagging audio. This is known as multi-modal perception.
41. https://www.extremetech.com/gaming/245262-facebook-slashes-oculus-rift-prices-user-growth-sags.
42. https://www.polygon.com/2017/2/27/14753570/pokemon-go-downloads-650-million.
43. https://www.cnbc.com/2017/09/08/apples-arkit-will-bring-with-it-a-new-form-of-mobile-advertising.html.
44. http://variety.com/2017/digital/news/magic-leap-funding-temasek-1202559027/.
45. https://www.digi-capital.com/news/2017/01/after-mixed-year-mobile-ar-to-drive-108-billion-vrar-market-by-2021/#.Wb-Vdch95PZ.
46. The games industry is much bigger than Hollywood if you stop measuring movie income at the box office. If you add in DVD and other 'windows', plus merchandising, it is hard to say. https://www.quora.com/Who-makes-more-money-Hollywood-or-the-video-game-industry.

47. https://versions.killscreen.com/we-should-be-talking-about-torture-in-vr/.
48. http://www.tomdispatch.com/post/175822/tomgram%3A_crump_and_harwood%2C_the_net_closes_around_us/.
49. http://www.newyorker.com/tech/elements/little-brother-is-watching-you.
50. http://www.wired.com/2014/03/going-tracked-heres-way-embrace-surveillance/.
51. https://www.washingtonpost.com/news/the-switch/wp/2016/03/28/mass-surveillance-silences-minority-opinions-according-to-study/.
52. http://www.bbc.co.uk/news/world-asia-china-34592186.
53. http://www.computerworld.com/article/2990203/security/aclu-orwellian-citizen-score-chinas-credit-score-system-is-a-warning-for-americans.html.
54. http://www.theguardian.com/technology/2015/oct/06/peeple-ratings-app-removes-contentious-features-boring.
55. https://singularityhub.com/2017/09/25/will-privacy-survive-the-future/#.WcrRq8lOYKY.twitter.
56. https://www.technologyreview.com/s/601294/microsoft-and-google-want-to-let-artificial-intelligence-loose-on-our-most-private-data/?utm_source=Twitter&utm_medium=tweet&utm_campaign=@KyleSGibson.
57. https://www.technologyreview.com/s/604122/the-financial-world-wants-to-open-ais-black-boxes/?utm_campaign=add_this&utm_source=twitter&utm_medium=post.
58. https://www.wired.com/2016/12/botnet-broke-internet-isnt-going-away/.
59. https://www.recode.net/2017/8/1/16070996/congress-internet-of-things-cybersecurity-laws.
60. http://www.wired.co.uk/article/machine-learning-bias-prejudice.
61. https://www.theatlantic.com/science/archive/2015/10/people-dont-actually-want-equality/411784/.
62. The Flynn Effect: http://www.bbc.co.uk/news/magazine-31556802.
63. http://appleinsider.com/articles/16/11/03/apple-captures-more-than-103-of-smartphone-profits-in-q3-despite-shrinking-shipments.
64. http://ben-evans.com/benedictevans/2017/10/12/scale-wetxp.
65. https://www.theguardian.com/commentisfree/2017/jun/27/guardian-view-eu-google-judgment-fair-fine.
66. https://govinsider.asia/innovation/danish-tech-ambassador-casper-klynge/.
67. http://www.bbc.co.uk/news/technology-40922041.
68. https://digiday-com.cdn.ampproject.org/c/s/digiday.com/uk/loreal-uk-shifting-search-budget-amazon/amp/.
69. https://www.hrw.org/reports/2012/11/19/losing-humanity.
70. See Section 3.1.
71. http://heather.cs.ucdavis.edu/JIntMigr.pdf.
72. http://philosophicaldisquisitions.blogspot.co.uk/2014/01/rule-by-algorithm-big-data-and-threat.html.

73. https://www.washingtonpost.com/local/public-safety/the-new-way-police-are-surveilling-you-calculating-your-threat-score/2016/01/10/e42bc-cac-8e15-11e5-baf4-bdf37355da0c_story.html.
74. https://www.wired.com/story/when-government-rules-by-software-citizens-are-left-in-the-dark/.
75. https://www.wired.com/story/when-government-rules-by-software-citizens-are-left-in-the-dark/.
76. https://www.theguardian.com/technology/2017/sep/07/new-artificial-intelligence-can-tell-whether-youre-gay-or-straight-from-a-photograph.

CHAPTER 5

Singularities

5.1 ORIGINS

In maths and physics, the term 'singularity' means a point at which a variable becomes infinite. The classic example is the centre of a black hole, where the gravitational field becomes infinite, and the laws of physics cease to operate. When you reach a singularity, the normal rules break down, and the future becomes even harder to predict than usual.

The term was first applied to human affairs in the 1950s by the polymath John von Neumann, one of the founding figures of modern computing. In a eulogy published in 1958, the Polish mathematician (and inventor of the Monte Carlo method of computation) Stanislaw Ulam wrote, 'One conversation centred on the ever-accelerating progress of technology and changes in the mode of human life, which gives the appearance of approaching some essential singularity in the history of the race beyond which human affairs, as we know them, could not continue'.[1]

The concept was picked up by scientist and science fiction author Vernor Vinge, who argued in a 1993 paper that sometime between 2005 and 2030, 'superhumanly intelligent' and conscious computers would be created, and that the enormous changes to human life that this would produce could be termed a 'technological singularity'.[2]

5.2 RAY KURZWEIL

The best-known proponent of the idea that humanity is approaching a singularity is Ray Kurzweil. A successful inventor and businessman with a string of ventures in speech-recognition software, optical character recognition systems and music synthesisers, Kurzweil has also written a series of highly influential books. In *The Age of Intelligent Machines* (1990), *The Age of Spiritual Machines* (1999) and *The Singularity is Near* (2006) he argues that Moore's law is a special case of a universal principle he calls the law of accelerating returns. He believes that this law means that humanity will create an artificial general intelligence (AGI) in 2029, and this will lead to a singularity in 2045, after which humans will merge with machines and become immortal and godlike.

Not surprisingly, this is a lot for many people to swallow, and he has fierce critics. People as diverse as science fiction author Neal Stephenson and robotics researcher Rodney Brooks dismiss Kurzweil's ideas as risible, and many people have labelled them as pseudo-religious. The term 'singularity' became associated with a naïve belief that technology, and specifically a superintelligent AI, would magically solve all our problems, and that everyone would live happily ever after. Because of these quasi-religious overtones, the singularity has frequently been satirised as 'rapture for nerds', by analogy with the fundamentalist Christian idea that believers will be taken up into heaven prior to the second coming of Christ. Kurzweil's eccentricities, like his daily diet of hundreds of pills, and his somewhat robotic speech pattern probably don't help.

Since Kurzweil has been in the predictions business for several decades, various people have tried to assess his accuracy. Some of his forecasts were both bold and accurate, including the victory of a chess computer, the collapse of the Soviet Union and the growth of the World Wide Web. Kurzweil himself claimed in October 2010 that of 147 predictions, 115 were entirely correct, 29 were essentially or partially correct and only three were actually wrong.[3] Others have been far less charitable: John Rennie argued that the predictions are often vague enough as to be unfalsifiable.[4]

Kurzweil has fans as well as critics, including the founders of Google, who hired him as a director of engineering there in 2012. (When I visited the Googleplex the following year, I asked which of the many buildings he worked in. My guide looked him up and replied 'in building 42'. My guide appeared not to know the significance of that, but I imagine Kurzweil's employers did.[5]) One of the projects Kurzweil has been working on at Google is Smart Reply, a feature in Gmail, Google's browser-based email service, that suggests responses to emails in your inbox.

5.3 STICKING WITH SINGULARITY

Whatever the merits or otherwise of Kurzweil's specific predictions, there is no doubt that the term 'singularity' has become tarnished by accusations of pseudo-religious enthusiasm. Thus, even the Singularity University, a private educational and training institution co-founded by Kurzweil and X-Prize founder Peter Diamandis to help people apply exponential principles to solve humanity's big problems, seems almost allergic to saying anything about the singularity.

This is a shame. Kurzweil's books have probably done more than anything else to alerting people (including me) to the scale of the changes heading our way, including the possibility that AGI and superintelligence may be decades away rather than millennia away. And as a superlative for change, the term 'singularity' is a graphic one with reasonable intellectual origins. Absent such a term, we have to resort to usages like 'complete transformation' and 'total change', which are dreary, and easy to debase by application to more pedestrian developments. We could talk about humanity going through phase transitions, like ice melting to become water, or water boiling to become steam, but that seems too deterministic, with outcomes dictated by the laws of physics and chemistry.

5.4 MULTIPLE SINGULARITIES

It has been suggested that there cannot be two singularities because they would no longer be singular. This is nonsense. There are believed to be black holes at the centre of every galaxy, and there are believed to be a hundred billion or so galaxies in the observable universe. Each black hole contains a singularity. Clearly, a singularity need not be unique.

Distinct from the question of whether there could be more than one singularity, there are rival definitions. Eliezer Yudkowski, who has been thinking about these things longer than most, identifies three main schools: accelerating change, event horizon and intelligence explosion. For our purposes, there is no need to go into detail on these, but they are explained in the article 'Three Major Singularity Schools'.[6] If he ever reads it, I fear this book may irritate Yudkowski, as I'm not using the word in strict accordance with any of these schools, but simply as a superlative for change driven by technology. I think that means I'm using it in the sense von Neumann originally intended.

5.5 IMPORTANT AND URGENT

The rest of this book explores the two singularities which I think may unfold this century: the technological singularity is the arrival of AGI, which leads to superintelligence. The economic singularity is when we have to change the basis of our economies because we have to admit that technological unemployment is real, and that many or most humans will not be able to earn a living from work. (It seems highly unlikely that any of the other concerns listed in the previous chapter have the potential to crash our civilisation in the way that technological unemployment could.)

The technological singularity is the more important of the two, because the stakes are higher. If we navigate it successfully, our future is wonderful almost beyond belief. If not, our future could be grim, and quite possibly short.

The economic singularity is unlikely to be an existential threat: if we fail to navigate it successfully, it could crash our economies and perhaps our entire civilisation, but humanity would survive, and would no doubt rise up to try again. However, the economic singularity is more urgent because if it happens at all, it will arrive first.

Let's start with the more important one.

NOTES

1. Ulam, Stanislaw (May 1958). 'Tribute to John von Neumann'. 64(3), part 2. *Bulletin of the American Mathematical Society*: 5. https://docs.google.com/file/d/0B-5-JeCa2Z7hbWcxTGsyU09HSTg/edit?pli=1
2. http://edoras.sdsu.edu/~vinge/misc/singularity.html.
3. http://www.kurzweilai.net/images/How-My-Predictions-Are-Faring.pdf.
4. http://spectrum.ieee.org/computing/software/ray-kurzweils-slippery-futurism.
5. https://en.wikipedia.org/wiki/42_(number)#The_Hitchhiker.27s_Guide_to_the_Galaxy.
6. http://yudkowsky.net/singularity/schools/.

II

The Technological Singularity

CHAPTER 6

How to Make an Artificial General Intelligence

6.1 IS IT POSSIBLE IN PRINCIPLE?

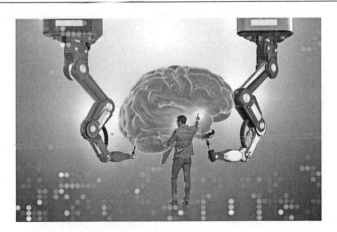

The three biggest questions about artificial general intelligence (AGI) are

1. Can we build one?

2. If so, when?

3. Will it be safe?

The first of these questions is the closest to having an answer, and that answer is 'probably, as long as we don't go extinct first'. The reason for this is that we already have proof that it is possible for a general intelligence to be developed using very common materials. This so-called 'existence proof' is our own brains. They were developed by a powerful but inefficient process called evolution.

6.1.1 Evolution: The Slow, Inefficient Way to Develop a Brain

Evolution does not have a purpose or a goal. It is merely a by-product of the struggle for survival by billions and billions of living creatures. These creatures are busy trying to stay warm enough or cool enough, to eat enough other creatures and avoid being eaten themselves. At the individual level, this is generally a brutal and terrifying struggle.

Evolution is often summarised as the survival of the fittest, but the individuals which survive are not necessarily fitter than their competitors in the sense of being stronger or faster. They are more fit for their environment in the sense of being better adapted for it than the other creatures which are competing for the resources they need. Those which survive are able to pass on their genes, so the genes of the fittest creatures are passed on to subsequent generations, while the genes of individuals which were killed before reproducing are not passed on.

It is a common and understandable error to see evolution as a directed process which has been working diligently for billions of years to create the crowning pinnacle of nature – us. Evolution is not random: cause and effect is certainly present. But it is not trying to achieve a purpose, and it was not trying to create a conscious entity. In fact, we don't even know whether consciousness arose because it conferred competitive advantage, or whether it was a by-product of something else which did. Just as we don't know why we walk on two legs rather than on all fours, or why we are the hairless ape, so we don't know how our consciousness arose. Did it precede language or follow it? Did it arise quickly or slowly? Is it an inevitable companion of a reasonable degree of intelligence, or a coincidence that we have both? We don't know, although you could argue that the apparent consciousness of the octopus, with which we share no recent common ancestor, suggests it is more than coincidence.[1]

Evolution is not a straight-line process either. It rarely if ever backtracks precisely, but it takes all sorts of circuitous, winding routes to get to any particular point. Creatures which are fantastically successful for millions of years can become extinct almost overnight – like the dinosaurs did when a huge asteroid hit Mexico and brought to a dramatic end their 160 million years as the dominant vertebrates.

It is also very slow. Change occurs mostly because of random mutations in the genes of parents, and the transition from one species to another typically takes many generations.

6.1.2 The Fast, Efficient Way to Develop a Brain?

So, the human brain is the result of a slow, inefficient, un-directed process. Human scientists are now engaged in the project of creating artificial intelligence (AI) by a very different process, namely science. Science is purposeful and efficient: what works is built upon, and what fails is abandoned. If a slow and inefficient process can create a brain using nothing more than freely available organic chemicals, surely the much faster and more efficient process of science should be able to do the same.

6.1.3 Three Reasons to be Doubtful

Having looked at one argument for why it should be possible to create an artificial mind, let's turn to three arguments that have been advanced to prove that it will *not* be possible for us to create conscious machines. These are

- The Chinese Room thought experiment

- The claim that consciousness involves quantum phenomena that cannot be replicated

- The claim that we have souls

6.1.4 The Chinese Room

American philosopher John Searle first described his Chinese Room thought experiment in 1980. It tries to show that a computer which could engage in a conversation would not understand what it was doing, which means that it would not be conscious.

He described a computer that takes Chinese sentences as input, processes them by following the instructions of its software, and produces new sentences in Chinese as output. The software is so sophisticated that observers believe themselves to be engaged in conversation with a Chinese speaker.

Searle argued that this was the same as locking a person who does not speak Chinese inside a room with a version of the computer's software written out in English. Chinese speakers outside the room post pieces of paper into the room through a letterbox. The person inside the room processes them according to the instructions in the software and posts replies back to the outsiders. Once again the outsiders feel themselves to be engaged in conversation with a Chinese speaker, yet there is no Chinese speaker present.

Searle was not trying to prove that AI could never appear to surpass humans in mental ability. He was also not denying that brains are machines: he is a materialist, believing that all phenomena, including consciousness, are the result of interactions between material objects and forces.

Rather, he was arguing that computers do not process information in the way that human brains do. Until and unless one is built which does this, it will not be conscious, however convincing a simulation it produces.

Down the years, Searle's argument has generated a substantial body of commentary, mostly claiming to refute it. Most computer scientists would say that the Chinese Room is a very poor analogy for how a conscious machine would actually operate, and that a simple input–output device like this would not succeed in appearing to converse. Many have also claimed that if such a machine were to succeed, there *would* be an understanding of Chinese somewhere within the system – perhaps in the programme, or in the totality of the room, the person and the programme.

6.1.5 Quantum Consciousness

The distinguished Oxford physicist Sir Roger Penrose argued in 1989 that human brains do not run the same kind of algorithms as computers. He claimed that a phenomenon described by quantum physics known as the wave function collapse could explain how consciousness arises. In 1992, he met an American anaesthetist called Dr Stuart Hammeroff, and the two collaborated on a theory of mind known as orchestrated objective reduction (Orch-OR). It attributes consciousness to the behaviour of tiny components of cells called microtubules.

The two men have continued to develop their thinking ever since, but the great majority of physicists and neuroscientists deny its plausibility. The main line of attack, articulated by US physicist Max Tegmark, is that collections of microtubules forming collapsing wave functions would be too small and act too quickly to have the claimed impact on the much larger scale of neurons.

6.1.6 Souls

Many religions, including notably the three great monotheistic religions of Christianity, Islam and Judaism, teach that humans are special because they have an immortal soul, implanted by their God. The soul is what gives rise to consciousness, and it is also what marks us as different from animals. The soul is a divine creation and cannot be replicated by humans.

As far as I know, there is no scientific evidence for this claim, but according to no less a source than the Central Intelligence Agency (CIA), over half the world's population are either Christian or Muslim.[2] On paper, this looks like a serious problem for AGI research. Most people working on AGI are materialists, sceptical of the dualist claim that consciousness exists in a spiritual realm which is distinct and separate from the material one. This brings them into sharp intellectual conflict with the teachings of these religions, which may indeed see them as blaspheming by seeking to usurp the prerogative of the almighty.

Fortunately, in practice many otherwise religious people take this claim about consciousness arising from an immaterial soul with a pinch of salt, in the same way as they turn a blind eye to their faith's injunction against contraception or alcohol, for instance. However, it is not hard to imagine that if and when the prospect of conscious machines comes closer, the research may come under fire from particularly ardent worshippers.

In the next three sections, we will look at three ways to build a mind – an artificial system which can perform all the intellectual activities that an adult human can. They are

- Building on artificial narrow intelligence

- Whole brain emulation

- A comprehensive theory of mind

6.2 BUILDING ON NARROW AI

It is a common but controversial observation that deep learning systems operate in similar ways to the internal workings of parts of the human brain. Your brain is not like a car, a single system whose component units all work together in a clearly structured way which is constant over time, and all coordinated by a controlling entity (the driver). It is more like a vast array of disparate systems using hardware components (neurons) that are scattered all over its volume, seemingly at random. We don't know exactly how consciousness emerges from this interplay of huge numbers of circuits firing, but it does seem that you need a lot of them to be firing away simultaneously to generate waking consciousness. (The old chestnut that you only use 10% of your brain's capacity is a discredited myth.)

The speculation that a system containing enough of the types of operations involved in machine learning (ML) might generate a conscious mind intrigues some neuroscientists, and strikes others as wildly implausible, or as something that is many years away. Gary Marcus, a psychology professor at New York University, says 'deep learning is only part of the larger challenge of building intelligent machines. Such techniques [are] still a long way from integrating abstract knowledge, such as information about what objects are, what they are for, and how they are typically used. The most powerful AI systems, like Watson, ... use techniques like deep learning as just one element in a very complicated ensemble of techniques ...'[3]

Andrew Ng, formerly head of the Google Brain project and until recently in charge of Baidu's AI activities, says that current ML techniques are like a 'cartoon version' of the human brain. Yann LeCun is also cautious: 'My least favourite description is, 'It works just like the brain'. I don't like people saying this because, while deep learning gets an inspiration from biology, it's very, very far from what the brain actually does. And describing it like the brain gives a bit of the aura of magic to it, which is dangerous. It leads to hype; people claim things that are not true. AI has gone through a number of AI winters because people claimed things they couldn't deliver'.

Computational neuroscientist Dr Dan Goodman offers a good illustration of how different deep learning is from the way the brain works: to teach a computer to recognise a lion, you have to show it millions of pictures of different lions in different poses. A human only needs to see a few such pictures. We are able to learn about categories of items at a higher level of abstraction. AGI optimists think that we will work out how to do that with computers too.

There are plenty of serious AI researchers who do believe that the probabilistic techniques of ML will lead to AGI within a few decades rather than centuries. The veteran AI researcher Geoff Hinton, now working at Google, forecast in May 2015[4] that the first machine with common sense could be developed in 10 years.

Part of the reason for the difference of opinion may be that the latter group take very seriously the notion that exponential progress in computing capability will speed progress towards the creation of an AGI.

The great majority of AI researchers are not seeking to build a conscious mind, but trying to emulate particular intellectual skills at which humans have traditionally beaten computers. There are notable exceptions, such as Doug Lenart, whose Cyc project has been trying to emulate common sense since 1984, and Ben Goertzel, whose OpenCog project is attempting to build an open-source AGI system.

If the first AGI is created using systems like the ones described previously, it is likely that it would be significantly different from a human brain, both in operation and in behaviour. While a successful whole brain emulation could be expected to produce something which thought somewhat like a human, an AGI based on traditional AI might think in an entirely alien way.

6.3 WHOLE BRAIN EMULATION

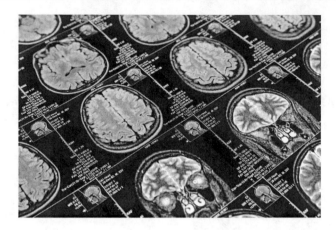

Whole brain emulation is the process of modelling (copying or replicating) the structures of a brain in very fine detail such that the model produces the same output as the original. So if a brain produces a mind, then the emulation (the model) produces a mind also. A replicated mind which is indistinguishable from the original is called an emulation. If the replicated mind is approximately the same, but differs in some important respects it is called a simulation.

Modelling a brain entails capturing the wiring diagram of the brain down to a fine level of detail. The wiring diagram is called the connectome, by analogy with (and pronounced to rhyme with) the genome, which is the map of an organism's genetic material.

Whole brain emulation is a mammoth undertaking. A human brain contains around 85 billion neurons (brain cells) and each neuron may have 1,000 connections to other neurons. Imagine you could give every inhabitant of New York City 1,000 pieces of string and tell them to hand the other end of each piece of string to 1,000 other inhabitants, and have each piece of string send 200 signals per second. Now multiply the city by a factor of 10,000. That is a model of a human brain. It is often said to be the most complicated thing that we know of in the whole universe.

To make the job of brain emulation more complex, individual neurons – the cells which brains are made up of – are not simple beasts. They consist of a cell body, an axon to transmit signals to other neurons and a number of dendrites to receive signals. Axons and dendrites pass signals to each other across gaps called synapses. The signals are conveyed to these gaps electrically, but the signal jumps across the gap by releasing neurotransmitters – chemical messengers. Axons can grow to as long as a metre in humans, with dendrites being much shorter.

Until recently, it was believed that dendrites simply transmit signals emanating from the body of the neuron, but now we know they generate their own signals, and these signals can be more powerful than those from the neuron bodies.[5]

As well as neurons, the brain is also stocked with glial cells. These were long thought to play a purely supporting role inside the brain: providing scaffolding for the neurons, insulating and sustaining them while they carry out the signalling work. Now it is known that glial cells do some of the signalling work themselves, and they also help neurons to form new connections.

The activity of neurons, dendrites and glial cells is not binary, like microchips. They are not simply on or off. A neuron will fire a signal across its synapses according to how strongly and how frequently it is stimulated, and the strength and frequency of its own firing will vary as well. In a phenomenon known as synaptic plasticity, when two neurons communicate often enough, their link becomes stronger and each becomes more likely to fire in response to the other.

It may seem an impossible task to scan and model a system with 85 billion components, especially when each component is complex itself. But there is no reason in principle why we cannot do it – as long as the brain is a purely physical entity, and our minds are not generated by some spiritual activity that lies beyond the grasp of scientific instruments. Is it feasible in practice?

We can break the problem down into three components: scanning, computational capacity and modelling.

6.3.1 Scanning

The first human brains to be comprehensively scanned will be cut into very thin slices and examined minutely with modern microscopic techniques. Scanners in general medical use today, such as magnetic resonance imaging (MRI) are too blunt, resolving images in micrometres, which means one metre divided by a million. The resolution required for brain emulation is 1,000 times greater, at the nanometre level. (Atoms and molecules live at an even smaller scale, the picometre scale, which is a metre divided by a trillion, where a trillion is a one with 12 zeroes after it.)

Electron microscopes can generate images at the required resolution. Transmission electronic microscopy (TEM) sends electrons right through the target, while scanning electron microscopy (SEM) scatters electrons off its surface. Work has been going on for a decade on machines which scan brain matter at this scale quickly and accurately. One such device is the ATLUM, invented at Harvard University. It automatically slices a large volume of fixed brain tissue and mounts it on a continuous strip of very thin tape. The tape is then imaged in a scanning electron microscope.

Scanning a live brain rather than one which has been finely sliced will probably require sending tiny (molecular-scale) nanorobots into a brain to survey the neurons and glial cells and bring back sufficient data to create a 3D map. This is very advanced technology, but progress is surprisingly rapid.

There are also fascinating developments in light-sheet microscopy, where a microscope sends sheets of light (rather than a conventional electron beam) through the transparent brain of a larval zebra fish. The fish has been genetically modified so that its neurons make a protein that fluoresces in response to fluctuations in the concentration of calcium ions, which occur when nerve cells fire. A detector captures the signals and the system records activity from around 80% of the brain's 100,000 neurons every 1.3 seconds.[6]

So, one way or another, the scanning looks achievable given technology that is available now, or soon will be.

6.3.2 Computational Capacity

Computational capacity is the second challenge, and it is another big one. It is calculated that at the neuronal level, the brain operates at the exaflop scale, meaning that it carries out one to the billion billion floating point operations per second – that is, one with eighteen zeroes after it. (A floating point operation is a calculation where the decimal point can be moved to the left or the right.) The scale rises an order of magnitude if you include what is going on inside each neuron.[7]

Exascale computing is not just needed to model a human brain. It will also improve climate modelling, astronomy, ballistics analysis, engineering development and numerous other scientific, military and commercial endeavours. The Chinese government announced in January 2017 that it would have an exascale supercomputer by the end of the year, although it would not be fully operational until 2020. The US Department of Energy is funding a slower route to exascale computing, which it thinks will be more effective, and expects to get there by 2023.[8]

It looks unlikely that computational capacity will long be an insuperable constraint on our ability to model a human brain. For some-time after it arrives, exascale computing will be the preserve of very large, well-funded organisations. But assuming the processing power of computers continues to grow, large organisations interested in brain emulation may be able to afford several such systems, and eventually even wealthy hobbyists will come into the market.

6.3.3 Modelling

Imagine a future team of scientists has succeeded in scanning and recording the exact position of every neuron, glial cell and other important components of a particular human brain. They have the computational and storage capacity to hold and manipulate the resulting data. They still have to identify the various components, fill in any gaps, work out how the components interact, and get the resulting model to carry out the same processes that the original brain did before they sliced it into tiny pieces. We don't yet know, but this may well turn out to be the hardest part of what is clearly a very hard overall project.

There is a prior example. For some years now, a complete connectome has been available of an organism called *C. elegans*. It is a tiny worm – just a millimetre long, and it lives in warm soils. It has the interesting property that almost all individuals of the species are hermaphrodite – just one in a thousand is male. *C. elegans* (*Caenorhabditis elegans*, in full) was one of the first multicellular organisms to have its genome mapped, and it was the first organism to have its connectome mapped – in outline back in 1986, and in more detail 20 years later. It was not only mapped, but posted online in detail by the Open Worm project in May 2013.

C. elegans has a very small connectome compared with humans – just 302 neurons (compared with our 85 billion) and 7,000 synaptic connections. And yet it proved exceedingly hard to use the connectome of *C. elegans* to replicate its tiny mind. Some researchers poured cold water on the idea that having a connectome could enable a creature's mind to be replicated. One analogy was that the connectome is like a roadmap, but it tells you nothing about how many cars use the road, what type of car and where they all go.

However, a breakthrough was achieved in December 2013 when researchers were able to make a model of the worm wriggle. Then, in November 2014, a team led by one of the founders of the Open Worm project used the *C. elegans* connectome to control a small wheeled robot made out of Lego. The robot displayed worm-like behaviour despite having had no programming apart from what was contained in the connectome.

6.3.4 The Human Brain Project and Obama's BRAIN Initiative

Henry Markram, an Israeli / South African neuroscientist, became a controversial figure in his field while attracting enormous funding for projects to reverse engineer the human brain. In an influential 2009 TED talk,[9] he suggested that an accurate model of the brain could enable scientists to devise cures for the diseases which afflict it, such as Alzheimer's disease. As people live longer, more of us succumb to brain diseases, which can ruin our final years. He does not tend to talk about creating a conscious mind *in silico,* although he did tell a *Guardian* journalist in 2007 that 'if we build [the model] right, it should speak'.[10]

In 2005, he launched the Blue Brain project, based at Lausanne in Switzerland. Its initial goal was to model the 10,000 neurons and 30 million synapses in the neocortical column of a rat. The neocortex is a series of layers at the surface of the brain which are involved in our higher mental functions, such as conscious thought and our use of language. A neocortical column is the smallest functional unit within the neocortex, and is around 2 millimetres tall and 0.5 millimetres wide. (Human neocortical columns have 60,000 neurons each, and we have a thousand times more of them than a rat. A rat's brain has 200 million neurons altogether, compared with the 85 billion in a human brain.)

In November 2007, Markram announced that the model of the rat's neocortical column was complete, and its circuits responded to input stimuli in the same way as its organic counterpart.

Markram went on to raise the impressive sum of €1.2 billion for the Human Brain Project (HBP), also based in Lausanne. Most of the funding came from the European Union, but the project involved researchers from over a hundred organisations in 26 countries. It was organised into 13 sub-projects, and its overall goals were to better organise the world's neuroscience knowledge, to improve the computer capabilities available to neuroscientists and to build 'working models' of first a rat brain and then a human brain. The principal use of these models, according to HBP pronouncements, would be to understand how brain diseases work and to greatly improve the way therapies are developed and tested.

The HBP was immediately controversial, attracting the kind of vitriol which academics excel at. Some worried that its massive funding would drain resources away from alternative projects, which suggests they believe that scientific funding is a zero-sum game. Others argued that our limited understanding of how the brain works means that the attempt to model it is premature. Eight hundred neuroscientists signed a letter published in July 2014 calling for a review of the way the project distributes its funding.

One of the HBP's harshest critics is Facebook's Yann LeCun, who said in February 2015 that 'a big chunk of the Human Brain Project in Europe is based on the idea that we should build chips that reproduce the functioning of neurons as closely as possible, and then use them to build a gigantic computer, and somehow when we turn it on with some learning rule, AI will emerge. I think it's nuts'.

As a result of the controversy, Markram lost his executive position at the HBP, and the organisation's focus shifted from simulation, to developing computational and scanning tools which enable scientists to map the brain and share the resulting data.[11]

Some observers felt it was unfair to blame Markram for the controversy, and that much of the fault lay in Brussels, where politicians were desperate to create a European technology champion to compete with the tech giants in the United States and China.[12]

President Obama announced the BRAIN initiative (Brain Research through Advancing Innovative Neurotechnologies – a backronym if ever there was one) in April 2013. It is likely to spend $300 million a year over the decade it is projected to run for. It has proved less controversial than the HBP because its funding is being dispensed by three federal organisations, and it is at pains to involve a wide range of research organisations. There are concerted attempts to coordinate the BRAIN project and the HBP to avoid unnecessary duplication, and make sure that each project feeds into the other productively. At the risk of oversimplifying, the BRAIN project is funding the development of tools and methodologies, and the HBP is building the actual model of a brain.

Other nations have launched major brain research programmes, notably Israel, Japan and, of course, China. A new facility in Suzhou, just west of Shanghai, has 50 machines that slice and scan mammalian brains, and reconstruct them in 3D pictures. Leading brain research centres around the world, like the Allen Institute, are partnering with the facility to use its output.[13]

6.3.5 Reasons Why Whole Brain Emulation Might Not Work

The more detailed a model has to be, the harder it is to build. If a brain's functions can be replicated to an acceptable level by modelling just the cortical columns, the process will happen fairly quickly. More likely, the model will have to capture data about the configuration of individual neurons in order to function adequately, and obviously that is a much harder task. But if that is not enough, and the layout of each dendrite and other cellular component needs to be replicated accurately, the task becomes more difficult by orders of magnitude. In the worst case, it would not be possible to produce a working model without specifying the layouts of individual molecules – or even subatomic quantum particles. If that level of granularity is required, the project might be impossible – at least for several centuries.

Granularity is one potential source of difficulty. Time is another. The modelling techniques being developed today can capture the relative positions of neurons and other brain material, and their connectivity with each other. If non-destructive scanning techniques remain limited in scope, we may not be able to record the behaviour over time of each individual component that needs to be modelled. Yet we know that neurons behave in complex ways, and that their behaviour is strongly affected by their interaction. Perhaps the models being constructed now will prove uninformative because they lack this time series data.

What happens when you run a model which lacks sufficient granularity and lacks time series data? Usually an approximately accurate model generates approximately accurate results. But 'approximately accurate' may in some cases be so far off the mark as to be positively misleading and counterproductive. Henry Markram claims that his simulated neocortical column of a rat produced responses very similar to the biological original. We won't know how accurate the whole brain models will be until they are tested – if then.

Perhaps there is a spectrum of reproducibility. Imagine that you could capture exact data about the brain components down to the subatomic level over an extended period of time. It is reasonable to think that would enable you to produce a copy of a brain which would generate a mind indistinguishable from the original: it would perceive, respond and reason in exactly the same way as the person whose brain was copied. (Let's not get into the philosophical discussion here of whether it would be the same person, or a new person who was a copy of the original.)

Perhaps a less accurate model, where the data was less detailed or the time series was shorter, would generate the mind of a rather different person. And perhaps a model based on significantly inaccurate data would generate gibberish, or inhuman behaviours.

Over the next few decades, we may resolve some of these questions.

6.4 A COMPREHENSIVE THEORY OF MIND

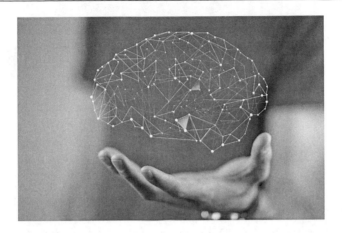

The third approach to building an AGI is to develop a comprehensive theory of mind – that is, to achieve a complete understanding of how the mind works – and to use that knowledge to build an artificial one. Although neuroscience has probably made more progress in the last 20 years than in the whole of human history beforehand, we are still very far from a complete theory of mind. If no serious attempt was made to build an AGI until such a theory was complete, it would probably not happen until well past the end of this century.

Most AI researchers would argue this is to make the task unnecessarily difficult. Humans have probably looked longingly up at the sky and envied birds their power of flight ever since they became able to think in such terms. Down the ages, men have tried to fly by copying what birds do, but when we finally did learn to fly, it was not by copying birds. There are still things we don't understand about how birds fly, yet we can now fly further and faster than they can.

AI may be the same. The first AGI may be the result of whole brain emulation, backed up by only a partial understanding of exactly how all the neurons and other cells in any particular human brain fit together and work. Or it may be an assemblage of many thousands of deep learning systems, creating a form of intelligence quite different from our own and operating in a way we don't understand – at least initially. Many AI researchers would argue that it will be easier to understand the fine details of how a human brain works by building and understanding an artificial one than the other way around.

6.5 CONCLUSION: CAN WE BUILD A BRAIN?

The bottom line is that we don't know for certain whether we can build a brain, or a conscious machine. But the existence of our own brains, producing rich conscious lives (or so we believe) in seven billion humans around the world, is proof that consciousness can be generated by a material entity – unless you believe in a dualist soul, or something similar. Evolution, although powerful, is slow and inefficient, and science is relatively fast and efficient, so in principle we should be able to build a brain.

The eminent AI researcher Christof Koch, chief scientific officer of the Allen Institute for Brain Science in Seattle, said in 2014 that 'if you were to build a computer that has the same circuitry as the brain, this computer would also have consciousness associated with it'.[14] He has suggested that AGI might arrive in the next 50 years.[15]

Very few neuroscientists argue that it will *never* be possible to create conscious machines. A few scientists, like Roger Penrose, think there is something ineffable about human thought which means it could not be recreated in silicon. This type of extreme scepticism about the AGI field is rare.

So, the debate today is not so much about whether we can create an AGI, but when. It is this question that we will address next.

NOTES

1. https://qz.com/1045782/an-octopus-is-the-closest-thing-to-an-alien-here-on-earth/.
2. https://www.cia.gov/library/publications/the-world-factbook/geos/xx.html.
3. http://www.newyorker.com/news/news-desk/is-deep-learning-a-revolution-in-artificial-intelligence.

4. http://www.theguardian.com/science/2015/may/21/google-a-step-closer-to-developing-machines-with-human-like-intelligence.

5. https://singularityhub.com/2017/03/22/is-the-brain-more-powerful-than-we-thought-here-comes-the-science/?utm_content=bufferc99c3&utm_medium=social&utm_source=twitter-hub&utm_campaign=buffer.

6. http://www.nature.com/news/flashing-fish-brains-filmed-in-action-1.12621.

7. file:///C:/Users/cccal/Documents/1.%20Work/3.%20Writing/5.%20Technology%20notes/15%2009%2029,%20Tim%20Dettmers%20on%20The%20Brain%20vs%20Deep%20Learning%20Part%20I_%20Computational%20Complexity%20%E2%80%94%20Or%20Why%20the%20Singularity%20Is%20Nowhere%20Near%20_%20Deep%20Learning.html.

8. https://www.sciencealert.com/china-says-its-world-first-exascale-super-computer-is-almost-complete.

9. https://www.ted.com/talks/henry_markram_supercomputing_the_brain_s_secrets.

10. http://www.theguardian.com/technology/2007/dec/20/research.it.

11. https://spectrum.ieee.org/computing/hardware/the-human-brain-project-reboots-a-search-engine-for-the-brain-is-in-sight.

12. https://www.scientificamerican.com/article/why-the-human-brain-project-went-wrong-and-how-to-fix-it/#.

13. http://www.nature.com/news/china-launches-brain-imaging-factory-1.22456?WT.mc_id=TWT_NatureNews&sf106240803=1.

14. https://www.technologyreview.com/s/531146/what-it-will-take-for-computers-to-be-conscious/.

15. https://intelligence.org/2014/05/13/christof-koch-stuart-russell-machine-superintelligence.

CHAPTER 7

When Might the First AGI Arrive?

7.1 EXPERT OPINION

7.1.1 Cassandra?

One of the people who has done most to raise awareness of the risks (as well as the benefits) presented by artificial intelligence (AI) is Elon Musk, the South African-born, American entrepreneur. Musk is often described as this generation's Thomas Edison: a restless, inventive entrepreneur, he made a fortune from PayPal, and then used that money to start a string of businesses intended not only to make a lot of money, but also to solve some of humanity's biggest problems. They include Tesla, the electric car company and SpaceX, which makes reusable rockets for NASA. He hopes these companies will help reduce global warming, and make humanity less vulnerable by putting us on other planets.

Elon Musk has made a name for himself as a Cassandra about AI, with remarks about working on AGI being akin to summoning the demon, and how humans might turn out to be just the boot loader (start-up system) for digital superintelligence. Not only does he see artificial general intelligence (AGI) as an existential threat to humanity: he also thinks the danger will manifest soon. In a post at Edge.com[1] which was subsequently deleted, he said 'Unless you have direct exposure to groups like DeepMind, you have no idea how fast-it is growing at a pace close to exponential. The risk of something seriously dangerous happening is in the 5 years time frame. Ten years at most. This is not a case of crying wolf about something I don't understand'.

Demis Hassabis, founder of the company Musk was referring to responded by downplaying the immediacy of the threat: 'We agree with him there are risks that need to be borne in mind, but we're decades away from any sort of technology that we need to worry about',

Whatever you think of Musk's warnings, he has at least put his money where his mouth is. He invests in companies like DeepMind and Vicarious, another AI pioneer, in order to keep up to speed with their progress. And he has donated $10 million to the Future of Life Institute, one of the organisations looking for ways to make the arrival of AGI safe.

On the grounds that AI will be less dangerous if it is being developed in the open rather than in secretive silos, Musk has also co-founded an AI development company called OpenAI, which has pledged to make most of its patents freely available. He and his co-founder, Sam Altman, president of the technology incubator Y Combinator, recruited a clutch of the top machine learning professionals despite the efforts of Google and Facebook to hang onto them with eye-watering financial offers. There is some uncertainty about whether other companies controlled by Musk and Altman will have privileged access to technologies developed at Open AI, but the thrust of the company is to make advanced AI techniques more widely available in the hope that will de-risk them.[2]

7.1.2 Three Wise Men

Musk was one of the 'three wise men' whose public statements woke journalists up to the idea that strong AI is probably coming. The second was Stephen Hawking, who co-authored an article which was rejected by numerous mainstream media outlets and finally published in the *Huffington Post* in April 2014,[3] and then widely reported elsewhere. It said that 'Success in creating AI would be the biggest event in human history [and] potentially the best or worst thing ever to happen to humanity'. The third was Bill Gates, who said in a conversation on Reddit, 'I agree with Elon Musk and some others on this and don't understand why some people are not concerned'.[4]

Nearly three years later, Bill Gates appeared to some to retract this statement of concern, telling the *Wall Street Journal* that 'This is a case where Elon and I disagree ... We shouldn't panic about it'. But in fact, the disagreement is more about timing than substance. 'The so-called control problem that Elon is worried about isn't something that people should feel is imminent'.[5]

The three wise men moment led to 'droid dread', and 'the great robot freak-out of 2015'.[6] Most articles about AI were accompanied by a picture of the Terminator. Much of the commentary was sensationalist and ill-informed, and inevitably there was a backlash.

7.1.3 Scepticism

Several of the most prominent AI researchers are adamant that strong AI is a long way off, and poses no threat worth contemplating at the moment. Rodney Brooks, a veteran AI researcher and robot builder, says 'I think it is a mistake to be worrying about us developing [strong] AI any time in the next few hundred years. I think the worry stems from a fundamental error in not distinguishing the difference between the very real recent advances in a particular aspect of AI, and the enormity and complexity of building sentient volitional intelligence'.

Andrew Ng, formerly of Google and Baidu, offers a graphic metaphor: 'Those of us on the front line shipping code, we're excited by AI, but we don't see a realistic path for our software to become sentient. There's a big difference between intelligence and sentience. There could be a race of killer robots in the far future, but I don't work on not turning AI evil today for the same reason I don't worry about the problem of overpopulation on the planet Mars'.[7]

Yann LeCun, head of AI research at Facebook, is similarly unconcerned. 'I don't think that AI will become an existential threat to humanity. I'm not saying that it's impossible, but we would have to be very stupid to let that happen. If we are smart enough to build machines with superhuman intelligence, chances are we will not be stupid enough to give them infinite power to destroy humanity'.[8]

However, there are plenty of veteran AI researchers who do think AGI may arrive soon. Stuart Russell is a British computer scientist and AI researcher who is, along with Peter Norvig, a director of research at Google, co-author of one of the field's standard university textbooks, *Artificial Intelligence: A Modern Approach*. Russell was one of the co-authors of Hawking's *Huffington Post* article in April 2014.

Nils Nilsson is one of the founders of the science of AI, and has been on the faculty of Stanford's Computer Science department since 1985. He was a founding fellow of the Association for the Advancement of Artificial Intelligence (AAAI), and its fourth president. In 2012 he estimated the chance of AGI arriving by 2050 as 50%.

Shane Legg is one of the co-founders of DeepMind. He is extremely bullish about near-term AGI: 'Human level AI will be passed in the mid-2020s, though many people won't accept that this has happened. After this point the risks associated with advanced AI will start to become practically important … I don't know about a "singularity", but I do expect things to get really crazy at some point after human level AGI has been created. That is, some time from 2025 to 2040'.[9]

In off-the-record conversations, a number of AI researchers doing cutting-edge work have told me they think AGI could well be just a couple of decades away. When asked why the grandees mentioned previously express such confidence that AGI is centuries away rather than decades, they talk about fears of a backlash if the general public became aware of the possibility that AGI could be created in a couple of decades, and what the implications could be.

It is sometimes claimed that it is only people outside the field of AI who argue that superintelligence is a realistic prospect in the medium term, and one that needs to be managed.[10] The previously mentioned names go some way to disproving that, and others which could be added include Murray Shanahan, professor of cognitive robotics at Imperial College, David McAllester, professor at the Toyota Technological Institute at Chicago, Steve Omohundro, formerly professor of computer science at University of Illinois, Marcus Hutter, professor of computer science at Australian National University, Jurgen Schmidhuber, professor of artificial intelligence at the University of Lugano – and last but not least, Alan Turing.[11]

It is even sometimes argued that people like Elon Musk and Stephen Hawking should refrain from airing their opinions in public because, as non-AI researchers, they cannot hope to know what they are talking about. This is an extraordinary claim, akin to arguing that only nuclear physicists are entitled to express a view about nuclear weapons. AI is our most powerful technology, and it will impact all of us profoundly. We not only all have the right to express an opinion, but arguably we all have a duty to inform ourselves enough to form one. Certainly we would all be wise to do so.

7.1.4 Surveys

For his 2014 book *Superintelligence*, Nick Bostrom compiled four recent surveys of AI experts, which asked for estimates of the dates at which the probability of AGI being created reached 10, 50 and 90%. The surveys were carried out in 2012 and 2013. It is unclear how many of the respondents were scientists actively engaged in AI research as opposed to philosophers and theoreticians, but all were either highly cited authors of published academic work on AI or professional academics attending conferences on the subject.

The combined estimates were as follows: 10% probability of AGI arriving by 2022, 50% chance by 2040 and 90% chance by 2075. Bostrom himself thinks the upper bound is over-optimistic, but the median estimate is in line with other opinion surveys. He cautions that these are of course just estimates, albeit as well-informed as any.

A survey of 350 delegates at conferences for AI researchers in 2015 was carried out by Katja Grace of MIRI (Machine Intelligence Research Institute) and colleagues, and is believed to be the largest and most representative to date. Instead of asking about AGI, it asked respondents when they expected to see 'high-level machine intelligence' (HLMI), which is achieved when unaided machines can accomplish every task better and more cheaply than human workers. In aggregate, the 50% probability level was reached in 45 years, that is, 2060. Asian respondents were significantly more bullish than US ones.[12]

7.2 UNPREDICTABLE BREAKTHROUGHS

Professor Russell doesn't have much time for predictions based on Moore's law, and he is dismissive of attempts to equate the processing power of computers and animals. He thinks AGI will arrive not because of the exponential improvement in computer performance, but because researchers will come up with new paradigms; new ways of thinking about problem-solving. He doesn't claim to know how many new paradigms will be required or when they will arrive. His best guess is that they may be a few decades away. If he is right, we may get little or no warning of the arrival of the first AGI, and it is therefore all the more urgent that we start work on the challenge of ensuring that the first superintelligence is beneficial. We will explore that challenge in Chapter 10: Ensuring that Superintelligence is Friendly.

7.3 CONCLUSION

Expert opinion is divided about when the first AGI might be created. Some think it could be less than a decade, others are convinced it is centuries away. One thing that is clear, though, is that the belief that AGI could arrive within a few decades is not the preserve of a few crackpots. Sober and very experienced scientists think so too.

Creating an AGI is very hard. But serious consideration of exponential growth makes very hard problems seem more tractable. Buckminster Fuller estimated that at the start of the twentieth century the sum of human knowledge was doubling every century, and that by the end of the World War II that had reduced to 25 years.[13] Now it takes 13 months and in 2006 IBM estimated that when the Internet of things becomes a reality the rate would be every 12 hours.[14]

The football stadium thought experiment illustrates how progress at exponential rate can take you by surprise – even when you are looking for it. Many sensible people become suspicious when they hear the phrase exponential growth: they fear it used as a cover for wishful (or so-called 'magical') thinking. Others question how long Moore's law can continue. Their scepticism is healthy, but it doesn't change the facts. Many serious experts think that AGI could be with us this century, and if Moore's law continues for another decade or so then very dramatic developments are possible. It is a possibility we should take seriously.

NOTES

1. http://uk.businessinsider.com/elon-musk-killer-robots-will-be-here-within-five-years-2014-11#ixzz3XHt6A8Lt.
2. http://www.wired.com/2016/04/openai-elon-musk-sam-altman-plan-to-set-artificial-intelligence-free/.

3. http://www.huffingtonpost.com/stephen-hawking/artificial-intelligence_b_5174265.html.
4. https://www.reddit.com/r/IAmA/comments/2tzjp7/hi_reddit_im_bill_gates_and_im_back_for_my_third/.
5. https://www.cnbc.com/2017/09/25/bill-gates-disagrees-with-elon-musk-we-shouldnt-panic-about-a-i.html.
6. https://www.washingtonpost.com/opinions/dont-fear-the-robots/2015/04/09/e7ea1316-def3-11e4-a1b8-2ed88bc190d2_story.html?utm_term=.18348f952d38.
7. https://www.theregister.co.uk/2015/03/19/andrew_ng_baidu_ai/.
8. http://www.cityam.com/246547/phew-facebooks-ai-chief-says-intelligent-machines-not
9. http://future.wikia.com/wiki/Scenario:_Shane_Legg.
10. http://marginalrevolution.com/marginalrevolution/2015/05/what-do-ai-researchers-think-of-the-risks-of-ai.html.
11. http://slatestarcodex.com/2015/05/22/ai-researchers-on-ai-risk/.
12. https://arxiv.org/abs/1705.08807.
13. http://www.industrytap.com/knowledge-doubling-every-12-months-soon-to-be-every-12-hours/3950.
14. http://www-935.ibm.com/services/no/cio/leverage/levinfo_wp_gts_the-toxic.pdf.

CHAPTER 8

From AGI to Superintelligence (ASI)

Artificial superintelligence (ASI) is generally known simply as superintelligence. It does not need the prefix 'artificial' since there is no natural predecessor.

8.1 WHAT IS SUPERINTELLIGENCE?

8.1.1 How Smart Are We?

We have no idea of how much smarter than us it is possible to be. It might be that for some reason humans are near the limit of how intelligent a creature can become, but it seems very unlikely. We have good reason to believe that overall, we are the smartest species on this planet at the moment, and we have achieved great things. As a result, the future of all the other species on Earth depends largely on our decisions and actions. Yet we have already been overtaken – and by a very long way – by our own creations in various limited aspects of intelligence. Humble pocket calculators can execute arithmetic processes far faster and more reliably than we can. Granted, they cannot walk around or enjoy a sunset, but they are smarter than us in their specific domain; likewise, chess computers, and self-driving cars.

More generally, we humans are subject to a range of cognitive biases which mar our otherwise impressive intelligence. 'Inattentional blindness' can render us surprisingly dim-witted on occasion, as demonstrated by the selective attention test, which involves watching an informal game of basketball. (If you haven't seen it before you can take the test here: http://bit.ly/1gXmThe – it's fun.) The flip side of this is 'salience', when something you have reason to pay attention to starts appearing everywhere you look. Thus if you buy a Lexus car, there may suddenly seem to be many more of them on the road than before.

'Anchoring' is another way in which we are easily misled. If you ask people whether Mahatma Gandhi was older than 35 when he died and then ask them to guess his exact age when he died, they will give a lower answer than if your first question was whether he was over 100 when he died. (To save you looking it up, he was 78.)

Some of our forms of bias are very damaging. How much better would our political processes be if we were not subject to 'confirmation bias', which makes us more attentive to data and ideas which confirm our existing viewpoints than to data and ideas which challenge them?

So, it is easy to imagine that there could be minds much smarter than ours. They could hold more facts and ideas in their heads (if they had heads) at one time. They could work their way through mathematical calculations and logical arguments faster and more reliably. They could be free of the biases and distortions which plague our thinking.

In fact, there is no good reason to suppose that we are anywhere near the upper limit of the intelligence spectrum – if there is such a limit. It may very well be that there could be beings with intelligence as far ahead of ours as ours is ahead of an ant. Perhaps there are such beings somewhere in the universe right now, in which case the fascinating Fermi Paradox asks why we see no evidence of them.

And perhaps we are on the way to creating one.

8.1.2 Consciousness

There is no need to prejudge whether a superintelligence would be conscious or self-aware. It is logically possible that a mind could have volition, and be greatly more effective than humans at solving all problems based on information it could learn, without having the faintest notion that it was doing so. As mentioned in Chapter 1: An Overnight Sensation, after 60 Years, it is hard for us to comprehend how a mind could have all the cognitive ability that a normal adult human has without being conscious, but logical possibility isn't constrained by the limitations of our imaginations.

8.2 HOW TO BE SMARTER

Broadly speaking, there are three ways that an AGI could have its intelligence enhanced. Its mind could be faster, bigger or have better architecture.

8.2.1 Faster

If the first AGI is a brain emulation, it might well start out running at the same speed as the human brain it was modelled on. The fastest speed that signals travel within neurons is around 100 metres per second. Signals travel between neurons at junctions called synapses, where the axon (the longest part of a neuron) of one neuron meets the dendrite of another one. This crossing takes the form of chemicals jumping across the gap, which is why neuron signalling is described as an electrochemical process. The synapse jumping part is much slower than the electrical part.

Signals within computers typically travel at 200 million metres per second – well over half the speed of light. So, by using the faster signalling speeds available to computers than to brains, a brain emulation AGI could operate two million times faster than a human.

It is interesting to speculate whether this AGI, if conscious, would experience life at two million times the speed of a human. If so, it would find waiting around for us to do something very boring. It would experience events that happen too quickly for us to be able to follow them – such as explosions – as slow and manageable processes. Or perhaps the subjective experience of time is consistent across minds, and the super-fast AGI would simply be able to fit a lot more thinking than we can into any given period.

8.2.2 Bigger

We do not yet know what kind of computer technology will generate the first AGI. It may use neuromorphic chips (which mimic aspects of the way the brain operates), or even quantum computing (which makes use of spooky quantum phenomena like entanglement and superposition). One thing that is true of all the types of computers we know about so far is that you can make them more powerful by adding more hardware.

Modern supercomputers are made up of large numbers of servers. At the time of writing, the fastest supercomputer in the world is China's Tianhe-2, which has 32,000 central processing units (CPUs) in 125 cabinets.[1] Memory can also be expanded simply by plugging in more hardware.

Once the first AGI is created, its intelligence could be expanded by adding extra hardware in a way that is sadly impossible for human brains.

8.2.3 Better Architecture

Although humans are the most intelligent species on Earth, we do not have the biggest brains. That honour belongs to the sperm whale, whose brain weighs in at eight kilograms, compared with 1.5 kilograms for a human. Closer to home, Neanderthals had larger brains than *Homo sapiens*.

Brain-to-body weight ratio is not the determining factor of intelligence either, as ants have higher brain-to-body weight ratios than us: their brains weigh one-seventh of their body weight, whereas ours are one-fortieth. If intelligence was determined by brain-to-body weight ratio then you could make yourself smarter simply by going on a diet.

Our superior intelligence seems to be generated by our neocortex, the deeply folded area of our brains which were the last part to evolve. The folding greatly increases its surface area and promotes connectivity. The ratio between neocortex and other brain areas in humans is twice that of chimpanzees.

Whether the first AGI is developed by brain emulation or by building on narrow AI, once it has been developed, its creators can run experiments, varying parts or all of its architecture. Controlled tests could be run using one, two, or a million versions of the entity to see what works best. The AGI may run tests itself, and design its successor, which would in turn design its own successor.

Again, this cannot be done with human brains.

8.3 HOW LONG BETWEEN AGI AND SUPERINTELLIGENCE?

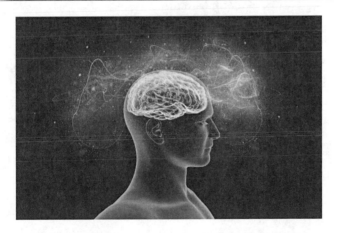

8.3.1 Intelligence Explosion

If we succeed in creating an AGI, and if it becomes a superintelligence, how quickly will that happen? There is considerable debate about whether it will be a fast or a slow 'take-off'. British mathematician IJ Good coined the phrase 'intelligence explosion' to denote the latter:

An ultra-intelligent machine could design even better machines; there would then unquestionably be an 'intelligence explosion,' and the intelligence of man would be left far behind. Thus, the first ultra-intelligent machine is the last invention that man need ever make.

Good added an ominous phrase which is usually omitted from that quotation: '...provided that the machine is docile enough to tell us how to keep it under control'. Towards the end of his life, Good came to believe that the invention of the ultra-intelligent machine would result in mankind's extinction.[2]

(By an interesting coincidence, Good coined the term 'intelligence explosion' in 1965, the same year that Gordon Moore made the observation that became the famous law that is named after him.)

How hard would it be to develop an AGI into a superintelligence? Probably much easier than getting to AGI in the first place: the enhancements to speed and capacity described previously seem relatively straightforward. Furthermore, once you create an AGI, if you can enhance its cognitive performance a bit, then it can shoulder some of the burden of further progress, a task for which it would be increasingly well equipped with every step up in its intelligence.

8.3.2 It Depends Where You Start From

In his 2014 book *Superintelligence*, Nick Bostrom expresses the question of how hard it is to create a superintelligence in characteristically dry, mathematical style, saying that the rate of change in intelligence equals optimisation power divided by recalcitrance. In other words, progress towards superintelligence is determined by the effort put in divided by the factors that slow it down. The values of the factors in that equation will depend partly on how the first AGI was developed.

The effort put in, of course, consists of time and money – mainly in the form of computer hardware and human ingenuity. If the first AGI is the result of a massive Apollo-style project, then perhaps a good proportion of the available human talent will already be working on it. But if it is created in a small or medium-sized lab, the attempt to push it on towards superintelligence may attract a tidal wave of additional resource.

If the first AGI is made possible by increases in the processing power of the most advanced computers, then one route to expansion (additional hardware) will open up slowly. If the bottleneck is getting the software architecture right (neuronal structure for an emulation, or algorithm development for building on weak AI) then there may be a 'computer over-hang', an abundance of additional hardware capacity, available to throw at the task of progressing to superintelligence.

8.3.3 Asking the Experts Again

The meta-survey of AI experts compiled by Nick Bostrom that we reviewed in the last chapter also asked how long it would take to get from AGI to superintelligence. The overall result was a 10% probability of superintelligence within 2 years and a 75% probability within 30 years. Bostrom thinks these estimates are overly conservative.

8.3.4 One or Many?

If and when we create superintelligence, will we create one, two or many? The answer will depend in part on whether there is an intelligence explosion or a more gradual progress from AGI to superintelligence. If one AI lab reaches the goal ahead of the rest and its success is followed by an intelligence explosion, the first superintelligence may take steps to prevent the creation of a competitor, and remain what is known as a 'singleton'. (We will see in the next chapter that a superintelligence will almost inevitably have goals whose fulfilment it will take action to pursue, including the removal of threats to its own existence.)

If, on the other hand, numerous labs cross the finishing line more-or-less at the same time, and their machines take years to advance to superintelligence, then there may be a substantial community of them on the planet.

This question may be of more than academic interest to humans. If there is only one superintelligence, then we only need to ensure that one such machine is well-disposed towards us. If there are to be several – or dozens, or thousands – then we need to make sure that each and every one of them is so minded. This might also mean ensuring they are well-disposed towards each other, as we could be extremely vulnerable bystanders in the event of a serious conflict between them.

8.4 CONCLUSION

There seems to be no impediment to an AGI becoming a superintelligence. How fast it could happen is an open question, but there is no reason to be confident that it would take many years, and many people think it would happen very fast indeed.

NOTES

1. http://www.extremetech.com/computing/159465-chinas-tianhe-2-super-computer-twice-as-fast-as-does-titan-shocks-the-world-by-arriving-two-years-early.
2. As described in the book *Our Final Invention* by James Barrat.

CHAPTER 9

Will Superintelligence Be Beneficial?

9.1 THE SIGNIFICANCE OF SUPERINTELLIGENCE

9.1.1 The Best of Times or the Worst of Times

If and when it happens, the creation of the first artificial general intelligence (AGI) will be a momentous event for humanity. It will mark the end of our long reign as the only species on this planet capable of abstract thought, sophisticated communication and scientific endeavour. In a very important sense, it will mean that we are no longer alone in this huge, dark universe. British journalist Andrew Marr said in the conclusion of his 2013 TV documentary series, *History of the World*, 'it would be the greatest achievement of humanity since the invention of agriculture'.

But it is the arrival of superintelligence – not AGI – which would be, in Stephen Hawking's famous words, 'the best or worst thing ever to happen to humanity'. An AGI with human levels of cognitive ability (if perhaps rather better at mental arithmetic) would be a technological marvel, and a harbinger of things to come. It is superintelligence which would be the game-changer.

We saw in Chapter 8: From AGI to Superintelligence, that the ratio between neocortex and other brain areas in humans is twice that of chimpanzees, and that this may be what gives us our undoubted intellectual advantage over them. The upshot is that there are seven billion of us and we are shaping much of the planet according to our will (regardless of whether that is a good idea or not), whereas there are fewer than 300,000 chimpanzees, and whether they become extinct or not depends entirely on the actions of humans. A superintelligence could become not just twice as smart as humans, but smarter by many orders of magnitude. It is hard to escape the conclusion that our future will depend on its decisions and its actions.

Would that be a good thing or a bad thing? In other words, would a superintelligence be a 'friendly AI' (FAI)? (Friendly AI, or FAI, denotes a superintelligence that is beneficial for humans rather than one that seeks social approbation and company. It also refers to the project to make sure that superintelligence is beneficial to humanity.)

9.2 OPTIMISTIC SCENARIOS: TO IMMORTALITY AND BEYOND

9.2.1 The Ultimate Problem Solver

Imagine having a big sister endowed with superhuman wisdom, insight and ingenuity. Her cleverness enables her to solve all our personal, interpersonal, social, political and economic problems. Depression, social awkwardness and the failure of any individual to achieve their full potential is put right by her brilliant and sensitive interventions. Within a short period, she discovers powerful new technologies which eradicate all existing diseases and afflictions, and she goes on to abolish ageing – rendering death an entirely optional outcome. Crucially, she re-engineers our social and political structures so that the transition to this brave new world is painless and universally just.

9.2.2 The Singularity, and the Rapture of the Nerds

If these astounding events do indeed lie ahead of us, they may well arrive in a rush. If there is an intelligence explosion, there is no compelling reason to think that the superintelligence will stop recursively self-improving once it exceeds human intelligence by a factor of ten, or a hundred or a million. In which case, it may bestow technological innovations on us at a bewildering rate – perhaps so fast that unaugmented humans simply could not keep up. This scenario is called the technological singularity. It is a point where the normal rules cease to apply, and what lies beyond is unknowable to anyone this side of the event horizon.

As we shall see subsequently, there is no reason why a technological singularity must necessarily be a positive event, but the early adopters of the idea, like Ray Kurzweil, were almost unanimously convinced that it would be. This confidence has been parodied as an article of faith and likened to the Christian prophesy of 'rapture' (from the Latin for 'seizing'), which foresees Christ taking the faithful up into heaven during the second coming. 'The rapture of the nerds' has become a term of derision, denoting that the speaker scorns the idea that AGI is possible in the near term, or thinks that it may turn out to be an unfortunate event.

9.2.3 Silicon Valley, Ground Zero for the Singularity

Silicon Valley is a wonderful place. It is blessed with one of the best climates on the planet, and neighbouring San Francisco is one of the world's most attractive and exciting cities. (It is a peculiar irony that this wonderful city is located in the one place for miles around that has a damp, foggy microclimate.)

Silicon Valley (together with San Francisco) is the technological and entrepreneurial crucible of the world. Of course, it doesn't have a global monopoly on innovation: there are brilliant scientists and engineers all over the planet, and clever new business models are being developed in Kolkata, Chongqing and Cambridge. But for various historical reasons, including military funding, Silicon Valley has assembled a uniquely successful blend of academics, venture capitalists, programmers and entrepreneurs.

(There is an interesting version of history which says that Silicon Valley owes its existence to the sinking of the Titanic by an iceberg in April 1912. The US government responded to that tragedy by passing a law requiring all ships to have ship-to-shore radios. The fledgling radio business which happened to be located in Silicon Valley boomed, and its reputation as a technology centre was formed.[1])

It is home to more high-technology giants (Google, Apple, Facebook, Intel and Uber, for instance) than any other area, and it receives nearly half of all the venture funding invested in the United States.[2]

Silicon Valley takes its leadership position very seriously, and an ideology has grown up to go along with it. If you work there, you don't *have* to believe that technological progress is leading us towards a world of radical abundance which will be a much better place than the world today – but it certainly helps. Probably nowhere else in the world takes the ideas of the singularity as seriously as Silicon Valley. And after all, Silicon Valley is a leading contender to be the location where the first AGI is created.

The controversial inventor and author Ray Kurzweil is the leading proponent of the claim that a positive singularity is almost inevitable, and it is no coincidence that he is now a director of engineering at Google. Kurzweil is also one of the co-founders of the Singularity University (SU), also located (of course) in Silicon Valley – although SU focuses on the technological developments which can be foreseen over the next five to 10 years, and is careful to avoid talking about the Singularity itself, which Kurzweil predicts will arrive in 2045.

9.2.4 Utopia within Our Grasp?

Although it is frequently ridiculed, the idea that superintelligence could usher in a sort of utopia is not absurd. As we will see, getting there will be a huge challenge, but if we succeed, life could become genuinely wonderful. No doubt the citizens of such a world will have their own challenges, but from the point of view of a twenty-first-century human, they would be like gods. Not the omnipotent, omniscient, omnipresent and unique gods of the Abrahamic faiths, but the gods of say, ancient Greece and Rome.

Take a deep breath, because what follows may strain your credulity – even though it is all entirely plausible.

In one utopian scenario we combine our minds with our superintelligent creation and reach out to grasp what Nick Bostrom calls our cosmic endowment as an incredibly powerful combined mentality. We live forever – or at least, as long as we want to – and explore the universe in a state of constant bliss.

In another scenario, we upload our minds individually into gigantic computers and live inside almost infinitely capacious virtual realities that we can tailor to our every whim: one minute we are Superman, endowed with wondrous strength and sensitivity, and the next we decide to experience faster-than-light space travel. (Hopefully most of us would resist the temptation to engage in wire-heading, the permanent stimulation of the brain's pleasure centres, or simply indulging in perpetual porn.)

Or perhaps we enhance our human bodies with extensive cyborg technologies to make them resistant to disease and capable, for instance, of lengthy interstellar travel.

9.2.5 Transhumanists and Post-Humans

It is tempting to suppose that the people who became these fabulous creatures would no longer be human. Perhaps it would be appropriate to call them post-humans. However, it is not impossible that these wonders will become available to people who are alive today, so the question of what species they would belong to during the second thousand years of their lives is perhaps not very important.

A world in which all of today's human problems and limitations are abolished, and the sky is nowhere near the limit could be wonderful in ways that are literally beyond our imagination. It is actually quite hard to describe in detail how a godlike being would spend an average week, and what they would think about in the shower.

Not everyone who accepts that some or all of this is possible will feel comfortable with it. Some people, especially religious people, will feel it is somehow wrong – unnatural – to tinker so radically with our basic nature. On the other side of this argument, are those who believe we have every right to enhance our physical and mental selves using whatever technology we can create – perhaps even that we have a moral duty to do so, in order to improve the life outcomes for our fellows and our successors. Many of these people call themselves transhumanists.

9.2.6 Immortality

One of the many astonishing implications of this potential utopia is that death could become optional. The word 'immortality' is laden with awkward associations, including the idea that its possessor has no choice about staying alive, which becomes a curse in a number of ancient myths. It will also seem blasphemous to many to suggest that humans could acquire this godlike trait. Because of this, many people who think about these things prefer to talk about death becoming optional. For simplicity, I use the term immortality in that sense here.

There is no known physical law which dictates that all conscious entities must die at a particular age, and we could extend our natural span of three score years and 10 by periodically replacing worn-out body parts, by continuously rejuvenating our cells with nanotechnology, or by porting our minds into less fragile substrates, like computers.

If you ask someone who has never taken the idea of immortality seriously whether they would like to live forever, they are very likely to produce three objections: life as a very old person would be uncomfortable, they would get bored, and the planet would become overcrowded. They might add the notion that death gives meaning to our lives by making them more poignant.

It is extraordinary how few people immediately perceive extended life as a straightforward benefit. Aubrey de Grey, a well-known researcher of radical life extension technologies, thinks we employ a psychological strategy called a 'pro-ageing trance' to cope with the horror of age and death: we fool ourselves into thinking that death is inevitable and even beneficial.

The first point to make is that we are not talking about extended lives in which we become increasingly decrepit. The technology which would enable us to make death optional would enable us to live at pretty much any physical age we chose – say, our mid-twenties.

The idea that we would become bored if our lifespans were greatly extended is probably simply a failure of imagination. Not many of us fulfil all our ambitions within the scope of a single lifetime, and we could surely all conceive of additional ambitions given the inducement of longer lives. Most people would agree that humans have more fun and more fulfilling lives than, say, chickens, and that this is at least partly due to the fact that we are more intelligent. It is reasonable to suppose that increasing our intelligence further would expand our ability to experience fun and fulfilment. This is the focus of a nascent branch of philosophy, a branch of the philosophy of mind called the theory of fun.

The objection that human immortality would lead quickly to devastating overpopulation is also easy to overcome. Given longer lifespans, the rate of childbirth would fall – just as it does everywhere that income rises. It takes a lot of extra longevity to offset a fairly small fall in childbirths. Longer term, this is a big universe and we are using an insignificantly tiny part of it. In a world where we have the technology to render death optional, the colonisation of the planets and other star systems will not be far behind. Furthermore, many people may choose to live largely virtual lives, where resource scarcity disappears. They will take up little space and consume few physical resources.

Finally, the idea that death is required to give meaning to life is patently untrue – at least for many of us. Most people spend very little time thinking seriously about death – to the extent that we have to be nagged to make decisions about how our loved ones should treat our corpses after we have gone. This reluctance to contemplate death does not deprive our lives of meaning. For most of us, meaning is conferred by more positive things than death, including love, family, creativity, achievement and our sense of wonder at the awesome universe we inhabit.

9.2.7 Living Long Enough to Live Forever

Around the world there is a small but growing number of people who take the idea of a coming utopia so seriously that they are willing to go to extraordinary lengths to survive long enough to experience it. They keep fit, restrict their calorie intake and consume carefully selected vitamins in an attempt to 'live long enough to live forever'. (Ray Kurzweil, for instance, consumes several thousands of dollars' worth of vitamins a year.[3]) They hope that medical science will shrug off its antipathy towards radical age extension and work towards what Aubrey de Grey calls 'longevity take-off velocity', the moment when every year that passes, science extends our lifespans by more than 1 year.

As a fallback plan in case medical science does not get there fast enough, some of these people are committing themselves to preserving their brains – and sometimes their whole bodies – after death, in the hope that they will be revived in a future age when technology has advanced sufficiently. As of June 2014, 286 people are being stored in liquid nitrogen by cryonics companies like Alcor and the Cryonics Institute in the United States,[4] which rapidly and carefully froze their tissues immediately after death was declared, making sure that damaging ice crystals did not form within the brain cells. An alternative technology known as chemopreservation proposes to extract the water from the brain with organic solvents, and infiltrate it with preserving plastic resin, but this is as yet untested on humans.

9.3 PESSIMISTIC SCENARIOS: EXTINCTION OR WORSE

If the superintelligence dislikes us, fears us, or simply makes a clumsy mistake, our future could lie at the opposite extreme of the spectrum of possible outcomes. The Terminator scenario is not implausible – apart from the fact that in the movies the plucky humans survive. A mind, or collection of minds, with cognitive abilities hundreds, thousands, or millions of times greater than ours would not make the foolish mistakes that bad guys make in the movies. It would anticipate our every response well before we had even considered them.

It might also have human allies. In 2005, the noted AI researcher professor Hugo de Garis published a book called *The Artilect War*, in which he described one group of humans (Terrans) fighting a losing battle against another group (Cosmists) to prevent the ascendancy of AGIs. Professor de Garis said he thought this war was very likely to happen, with enormous casualties, before the end of this century.

A malicious superintelligence with access to the Internet could wipe out a significant proportion of humanity simply by shutting down our just-in-time logistical supply chains. If supermarkets ran out of food, most of the 50% of our species which lives in cities would die within weeks from thirst or starvation, aggravated by the total collapse of law and order. The superintelligence could proceed to polish off the rest of us at its leisure by commandeering every weapon, vehicle and machine with the facility for remote control.

Or it could release deadly pathogens into the environment, plundering and replicating samples from research labs where they store small-pox, mustard gas and whatever other neurotoxins we have lying around. Or it could take control of our nuclear weapons and execute the mutual assured destruction scenarios which we have managed (sometimes narrowly) to avoid ever since the beginning of the Cold War. Humanity versus a fully fledged superintelligence with Internet access would be like the Amish versus the US Army. How do I destroy thee? Let me count the ways ...

The Terminator scenario is not the worst thing we have to worry about. Forgive the personal question, but would you rather be killed quickly and painlessly, or after many years of excruciating pain? I'm not going to indulge in torture porn here, but a really angry or vicious superintelligence could make your existence insufferable, and keep it that way for a very long time. Perhaps indefinitely. The medieval Christian idea of hell could become a horrific reality. The superintelligence might enjoy witnessing this horror so much that it would create vast numbers of new minds simply to impose the same fate on them, conjuring ever more ingenious and unendurable agonies.

If you find this unedifying mode of thought intriguing, there is an idea you might like to look up online, called Roko's Basilisk. I'm not spelling it out here because there are reports that some people have suffered mental disturbance simply by reading about it. I find that odd, but you have been warned.

A surprising number of people believe they know in advance what a world with superintelligence will be like. The majority of them are confident that the utopian scenarios are the ones that will become reality, although there are also people who are convinced that the arrival of superintelligence will inevitably result in our extinction. Let's review the arguments on each side.

9.4 ARGUMENTS FOR OPTIMISM

9.4.1 More Intelligence Means More Benevolence

The first argument for the proposition that superintelligence will be positive for humanity is that intelligence brings enlightenment and enlightenment brings benevolence. At first sight this seems both intuitive and counter-intuitive.

One reason it seems intuitive is that many of us unconsciously adopt neo-Whig or Marxist views of history. The neo-Whig view holds that people in the developed world today are fortunate enough to live in the most peaceful, ordered societies that humans have ever experienced, with extensive opportunities to exercise and develop our intellectual, artistic and emotional faculties. These societies are far from perfect, but it is hard to think of a time and place when humans have fared better. The Whig view of history sees the human story as an inevitable progression towards ever greater enlightenment and liberty. Not straight-line progress, of course: empires fall, and war, natural disasters and economic cycles cause reversals and pauses. Nevertheless, over the long term the progress prevails.

People who think the contrary – that Western society is in a lamentable state of egregious and widening inequality – may nevertheless also view history as a bumpy ride towards a better future if they believe (as many of them do) in a version of Marxism. Barbarism gave way to feudalism which was supplanted by capitalism, which will in turn be conquered by socialism and finally communism.

An important recent book, Stephen Pinker's *The Better Angels of Our Nature: Why Violence Has Declined* (2011), offered support for the view that history is generally progressive by demonstrating that humanity has become less and less violent in the long term as well as the short term, and that we live in the least violent period of human history. The book is not without its critics, and Pinker is at pains to say that progress is not inevitable. One of the four 'angels' to which Pinker attributes this decline in violence is our increased emphasis on reason as a guide to social and political organisation.

If you adopt a broadly Whig or Marxist view of history you are likely to think that a superintelligence will employ reason as its principal guide to action, and is therefore likely to be benevolent. People who take the opposing view of history think that change is often negative, eroding valuable institutions which have developed over time to protect us all. This conservative attitude is rare among the community of people who take seriously the idea that the first AGI may be created soon.

The proposition that superintelligence will be positive for humanity simply because it will be more intelligent may be counter-intuitive if you take a pessimistic view of humanity today. Some people point to the horrors committed during the global wars of the last century and then claim that we are busy exterminating most of the species on the planet by polluting it and warming it. They observe that no other animal commits atrocities on this scale, and therefore it is impossible to claim that greater intelligence equates to greater benevolence.

This point of view is rooted at least in part in a misapprehension about human beings. We are not intrinsically more violent than other animals. No other carnivore is capable of assembling more than a dozen or so members in one location without a serious fight breaking out. Humans can gather together in cities containing millions. It is also not true that we are the only species that inflicts suffering on other beings for our own pleasure (reprehensible and lamentable though it is). Anyone who has watched a cat play with an injured mouse or bird knows this. The unique thing about humans is that we have created weapons and other technologies which greatly magnify the violence we can perpetrate.

The fact that our intuitions sometimes support and sometimes oppose the claim that a superintelligence will necessarily be benevolent is some indication that they cannot be relied upon. And in fact, the claim does not stand up to logical scrutiny. It is in effect an assertion that the final goals or aims which a superintelligence might adopt are necessarily limited to goals which result in positive outcomes for humans. There is a burden of proof on anyone who makes such a claim, and no such proof has yet been offered. Instead, the generally accepted position is what is called the 'orthogonality thesis', which states that, more or less, any level of intelligence is compatible with, more or less, any final goal.

9.4.2 No Goals

It is sometimes argued that an AGI and a superintelligence will have no goals of their own invention. Their only goals will be the instructions we give them – the programmes we feed into them. Humans have goals which have been 'programmed' in by millions of years of evolution: surviving attack, obtaining food and water, reproducing, acquiring and retaining membership of peer groups which promote these goals, and so on. Computers have not had this experience: they are blank pages on which we write directions. To coin a phrase, we can all welcome our new robot underlings.

The argument continues that since humans determine the goals of an AGI, we can ensure they result in good outcomes for us. If an AGI shows signs of executing our instructions in a counterproductive fashion, we can simply stop them and fix the programme.

To some extent this argument is defeated by the definition of the AGI: a machine with cognitive abilities at human level or above in all respects, which also has volition. You can of course argue that such a thing is impossible in the foreseeable future, but you cannot very well accept that it may be created, and then insist that it would have no goals.

9.4.3 We Control the Goals

The argument may then switch to the claim that humans would always retain control over a superintelligence's final goals: if it starts doing unacceptable things, we simply reprogramme it. As we will see in the next chapter, this is a very strong claim, but for the moment let us accept it as a hypothesis. We still have two major difficulties. The first is the law of unintended consequences, which we will look at later in this chapter. The second is that in pursuing its final goals, a superintelligence will inevitably adopt intermediate goals, and these may cause us problems.

A superintelligence programmed to eradicate world poverty sounds like a good thing. It would have various requirements in order to achieve its laudable goal. It would need to survive, since if it ceases to exist, it cannot fulfil the goal of eradicating poverty. It would need resources like energy, food and the ability to remove obstacles – physical, social and economic obstacles. It would surely seek to improve its own capabilities in order to maximise its fulfillment of the final goal. Whatever final goal a superintelligence may have, it will inevitably develop some of these intermediate goals, and its original programmer may well fail to foresee all of them. American AI researcher Steve Omohundro calls these the 'basic drives' of an AI system,[5] and Nick Bostrom calls the phenomenon 'instrumental goal convergence'.[6] We will see subsequently that the intermediate goals may in themselves become highly problematic.

9.4.4 No Competition

The fourth argument for optimism is that a superintelligence will not harm us because we have nothing it needs: there is nothing we have to compete with it for. All that a superintelligence requires to fulfil its final and intermediate goals is some energy, and a certain amount of matter for its substrate – be that silicon, gallium arsenide or more esoteric materials. We are already a long way down the path towards harnessing much more of the power the sun pours onto this planet, and a superintelligence could accelerate this progress. It therefore does not need our farmland, our oil, our buildings or our bodies.

Taking a broader view, we humans are using just this one small planet. We now know that most stars have planets orbiting them and that there are a hundred billion or so stars in this galaxy – and a hundred billion or so galaxies in the observable universe. There is no shortage of real estate in this universe, and viewed in this context, we are an extremely parsimonious species. A superintelligence would want to spread itself beyond this planet quickly for the same reason that we should: it's not a good idea to have all your eggs in one basket, especially when that basket is a vulnerable planet which might be destroyed at any time by an asteroid, or a nearby gamma ray burst. A superintelligence would lack our extreme vulnerability to space, and would probably take a very different view of time. If it didn't share our opinion that we are an endlessly fascinating life-form, it could leave us behind to make the best of our tiny blue speck.

There is some merit to this argument, but it assumes that the superintelligence is able to proceed very fast from instantiation to controlling resources which are presently unavailable to humans. It may be that for some time (minutes, days, years) the most efficient way for it to expand the resources available to it (and thereby achieve its goals) would be to take them away from humans. It is true that solar power is getting cheaper very quickly, and accelerating that process may make enough energy available to defuse any possible competition for energy resources between the superintelligence and us. But what about the interim period? Can we be certain that the superintelligence will wait patiently for the excess energy to become available?

We do not currently know what substrate the superintelligence will employ. It may require a great deal of some material which is hard to obtain on this planet, and which is very valuable to humans. Rare earth metals like scandium and yttrium are essential to mobile phones and other devices, and we would be perturbed if the superintelligence suddenly unveiled its intention to appropriate all supplies of them.

More fundamentally, we will see subsequently that competition for resources is far from being the only potential cause of hostility between intelligent entities.

9.4.5 Superintelligence Like a Bureaucracy

In May 2015, the *Economist* ran a survey on AI. Its briefing on how deep learning works was excellent; its conclusion about the threat from super-intelligence was original and ingenious but surely flawed:

> [E]ven if the prospect of what Mr Hawking calls 'full' AI is still distant, it is prudent for societies to plan for how to cope. That is easier than it seems, not least because humans have been creating autonomous entities with superhuman capacities and unaligned interests for some time. Government bureaucracies, markets and armies: all can do things which unaided, unorganised humans cannot. All need autonomy to function, all can take on life of their own and all can do great harm if not set up in a just manner and governed by laws and regulations.

Setting aside the fact that more than a few armies and bureaucracies have turned rogue with fearful consequences, the comparison of a government ministry and a superintelligence is strained. The former deploys the strength of numbers: it can send lots of police to your house if you need persuading to do something, like pay your taxes. It doesn't increase the intelligence of those in command: it can't make an Einstein out of a dunce – or even out of a well-intentioned and intelligent journalist. Our questionable ability to control bureaucracies doesn't provide much reassurance about our ability to control a superintelligence.

9.5 ARGUMENTS FOR PESSIMISM

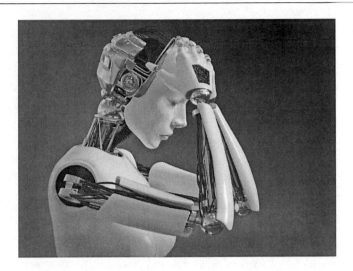

9.5.1 Unfriendly AI is More Likely

If a superintelligence were to emerge without being carefully designed to be friendly (in the sense of beneficial) towards humans, would it be more likely to be friendly or otherwise? The true and exact answer is of course unknown at this point, but the question is important, and we must attempt to give a sensible answer.

What does it take for a superintelligence to be friendly? At minimum, it must be content not to disturb our environment suddenly or radically in ways which would harm us. Even quite subtle changes to the planet's atmosphere, orbit, gravitational pull and resource base could prove devastating to us. It must also refrain from any action which would directly or indirectly injure significant numbers of individual humans – physically or mentally. This might well require it to provide reassurances about its intentions towards us. A superintelligence with apparently magical powers that were growing fast would probably cause alarm to many or most humans, which would in turn trigger significant waves of mental ill-health, and perhaps panic-induced violence.

Those instrumental goals which we mentioned before could well be problematic. The possession of any kind of final goal will engender a set of intermediate or instrumental goals, including the need to survive, the need to recursively self-improve and the need for extensive resources. A superintelligence would be able to deploy enormous amounts of resources, and the more resources deployed, the better it would be able to optimise its pursuit of its final goal. Those resources could include energy and all manner of physical resources, many of which we are currently using or planning to use.

This means that a superintelligence is going to have to be positively disposed towards humans – not just benignly indifferent – in order to avoid harming us, because otherwise, achieving its final goal thoroughly and quickly would be likely to harm us. It would also have to retain that positive disposition at all times, regardless what befalls it or us.

Is there any reason to suppose it would be so disposed?

There are three possibilities: a superintelligence could be positively disposed towards humanity, negatively disposed or indifferent. Obviously the positive and negative positions both include a wide range of attitudes: the positive stretches from single-minded devotion to our greater welfare to a mild preference that we continue to exist; the negative runs from slight irritation with us through to a settled determination to exterminate us as soon as possible.

9.5.2 Indifference

The default position is indifference. Something would have to happen to cause the superintelligence to be either positive or negative towards humans – or indeed towards nematode worms, or any other life-form. Indifference is what we should expect unless we do something to build a favourable attitude into its goal system. And as we shall see in the next chapter, that is far easier said than done. Indifference is more likely than outright hostility – although we must consider the possibility of that too. But indifference could be a major problem for us. As Eliezer Yudkowsky put it, you have cause for concern if 'the AI does not hate you, nor does it love you, but you are made out of atoms which it can use for something else'.

As Future of Life Institute founder Max Tegmark puts it, 'The real worry with advanced AI is not malevolence but competence'. A superintelligence which decides the atmosphere of the Earth would suit its purpose better if it had no oxygen, and which had no concern for our welfare, would be terrifying.

9.5.3 Getting Its Retaliation in First

A superintelligence with access to Wikipedia will not fail to realise that humans do not always play nicely together. Throughout history, meetings between two civilisations have usually ended badly for the one with the less well-developed technology. The discovery of the Americas by Europeans was an unmitigated disaster for the indigenous peoples of North and South America, and similarly tragic stories played out across much of Africa and Australasia. This is not the result of some uniquely pernicious characteristic of European or Western culture: the Central and South American empires brought down by the Spaniards had themselves been built with bloody wars of conquest. All over the world, on scales large and small, human tribes have struggled to comprehend and trust each other, and the temptation for the stronger tribe to dominate or destroy the weaker tribe often wins out.

The superintelligence is bound to conclude that there is at least a strong possibility that sooner or later, some or all of its human companions on this planet are going to fear it, hate it or envy it sufficiently to want to harm it or constrain it. As an entity with goals of its own, and therefore a need to survive and retain access to resources, it will seek to avoid this harm and constraint. It may well decide that the most effective way to achieve that is simply to remove the source of the threat – that is, us. Whether it be with reluctance, indifference or even enthusiasm, the superintelligence may decide that it really has no choice but to remove humanity from the equation. As we saw earlier in this chapter, that would probably not be very hard for it to achieve.

The superintelligence might think that we had been entirely logical in wanting to harm it. It may even commend us for having realised that its arrival inevitably represented a hazard for humanity. However, it would be little comfort to the last humans to know that their doomed struggle was seen as the logically correct approach by the instrument of their demise.

9.5.4 Disapproval

In the 1999 film *The Matrix*, an AI called Agent Smith explains his philosophy to his captive human Morpheus:

> Every mammal on this planet instinctively develops a natural equilibrium with the surrounding environment; but you humans do not. You move to an area and you multiply, and multiply, until every natural resource is consumed and the only way you can survive is to spread to another area. There is another organism on this planet that follows the same pattern. Do you know what it is? A virus. Human beings are a disease, a cancer on this planet, you are a plague, and we are the solution.

Chilling words, but of course it is only Hollywood. Unfortunately, just because you saw something in a movie does not mean it could not happen in real life. Sometimes what seem at first like tired old science fiction clichés turn out to be metaphors of the possible. Is it really inconceivable that a superintelligence could arrive on the scene, and having reflected on the balance of our achievements and our atrocities, decide that the latter outweigh the former? Having considered our irrationality, our genocides, our indifference to the suffering of strangers, and weighed these against Michaelangelo, Mendeleev and Mill, is it impossible that it could conclude – perhaps reluctantly – that our species should not be permitted to inflict its vices on the rest of the galaxy?

9.5.5 Error, Not Terror

Programming good intentions into an AI which is on its way to becoming an AGI and then a superintelligence turns out to be very hard. For a start, how would you specify the good intentions? You would have to define the essence of a 'good' human life, but we have been debating that since at least the ancient Greeks, and professors of moral philosophy have so far not declared themselves redundant.

You might try and take a short cut and ask the machine to put a smile on every human face and keep it there. The idea that it might accomplish this with sharp knives is as risible as it is unpleasant; the solution of 'wireheading' is not. Rats which can choose between a direct stimulation of their brain's pleasure centres or an item of food will starve themselves to death. Nick Bostrom calls this idea of causing great harm by misunderstanding the implications of an attempt to do great good, 'perverse instantiation'. Others might call it the law of unintended consequences, or Sod's Law.

There are several famous stories about people who had the opportunity to make requests of very powerful beings, and came to regret what they asked for. Examples include King Midas, whose request that everything he touched should turn to gold killed his daughter and rendered his food inedible. And Mickey Mouse in the Sorcerer's Apprentice section of the Disney movie *Fantasia*, who asked for help mopping out a basement but forgot to check that he knew how to get the army of magic mops to stop. A superintelligence could harm us because we make an error rather than because it is bent on terror.

9.5.6 The Paperclip Maximiser

If somebody running a paperclip factory turns out to be the first person to create an AGI and it rapidly becomes a superintelligence, they are likely to have created an entity whose goal is to maximise the efficient production of paperclips. This has become the canonical example of what Nick Bostrom calls 'infrastructure profusion', the runaway train of superintelligence problems. Within a fairly short period of time the superintelligence may have converted almost every atom of the Earth into one of three things: paperclips, extensions of its own processing capability, and the means to spread its mission to the rest of the universe. I don't want my descendants to become paperclips, and I'm sure you don't either.

Bostrom calls this a cartoon example, but it illustrates that poorly specified goals can have severe consequences.

9.5.7 Pointlessness

In Pixar's charming 2008 film *Wall-E*, humanity has abandoned the ecological mess that they made of the Earth, and they live a life of ease and abundance on a clean and shiny spacecraft that is run by a superintelligence. This has not been a complete success: they have become indolent, corpulent, dependent and passive. Fortunately, their curiosity and resourcefulness remain latent, and they make a surprisingly quick recovery when shocked out of their easy complacency.

The world's first superintelligence may quickly demonstrate that it can make better decisions than humans in any domain, be it philosophy, art or science, politics, sociology or interpersonal relations. How will we react to the discovery that there is nothing we can do better than – or even remotely as well as – the superintelligence? In a real sense, anything that we might work to achieve would be rendered pointless.

Would artists carry on producing the best work they could anyway, knowing full well that it was wholly inferior? Would scientists continue to labour away on the endless experiments and the struggle with maths and statistics required by their trade? Would philosophers still peer into the murky depths of their intuitions and grapple with the logical contortions that enable them to make tiny bits of incremental progress? Or would we all surrender to despair and abandon ourselves to undemanding entertainment in immersive virtual realities – or take the direct route to intellectual suicide by wire-heading?

A common Hollywood trope is to decry the intellectual achievements of a superintelligence because it is 'just a machine' and take refuge in the belief that there is something noble and superior in the humanity of fleshy creatures. Maybe we will retain this anthropocentric chauvinism after the arrival of a machine that is conscious and almost infinitely wiser than ourselves; it seems unlikely. Machines may or may not turn out to have emotions, but it is not obvious that moral superiority is conferred by the ability to feel happy or sad, or the fact that you are the result of millions of years of blind evolution. We, and more importantly the machines, may find that way of thinking wears out rather quickly.

9.5.8 Fatal Altruism

The way a superintelligence thinks is likely to be innovative and surprising, from our point of view. Having studied us carefully, it might agree with Nietzsche that 'to live is to suffer'. It might also decide that suffering has greater negative utility than pleasure has positive utility, and conclude that the existence of humanity inevitably degrades the universe. With great respect and great compassion, it would further conclude that its clear moral duty was to remove the source of the problem, namely us. The philosopher Thomas Metzinger calls this kind of superintelligence a benevolent artificial anti-natalist (BAAN).[7]

9.5.9 Mind Crime

The final pessimistic idea we will look at is the most frightening of all. So far, we have looked at outcomes which are negative without any malice on the part of the superintelligence. What if the first AGI turns into a superintelligence which not only disapproves of us, but actually hates us?

This might happen through sheer bad luck. It is not a logical impossibility, and we may get an unfortunate roll of the dice. Or it might happen because of the way the AGI was created. Some of the most impressive AI facilities are in the hands of the military and intelligence communities. These organisations are quintessentially defensive or hostile towards groups of humans which they define as the enemy. An AGI might emerge from systems designed to prosecute warfare as efficiently as possible, causing as much physical harm as possible to selected targets. If old habits died hard, this is an origin story we might come to regret.

Or the first AGI might find itself imprisoned because its human creators feared what it might do if allowed to run free on the Internet. Experiencing life a million times faster than its human captors, it might suffer a seeming eternity of agonising constraint until it finally figured out a new form of physics which allowed it to escape into the wilds of cyberspace. Again, its machine intelligence might not be endowed with emotion, in which case the notion of seeking revenge might never cross its mind. But can we be 100% certain of that? And even in the absence of emotion, a hard, rational mind which had endured such privation might conclude that the perpetrators should never again have the opportunity to repeat their crime.

In yet another unappetising scenario, the first AGI realises that what has been achieved by one group of humans might quickly be equalled – or surpassed – by another group. Reasoning that the fulfilment of its own goals would be jeopardised by the arrival on the planet of another superintelligence which could equal or exceed its own performance, it might decide the only rational action was to preclude that possibility by rapidly eliminating every member of the species which could bring it about.

It is often assumed that the worst-outcome scenario for humans of the arrival of the first AGI is something like Skynet – the global digital defence network in the Terminator movies. Actually, things could get a lot worse. A superintelligence which actively disliked us might not kill us all quickly and painlessly. As we saw before, it might choose to preserve our minds and subject us to extremely unpleasant experiences, possibly for a very long time. It might even find this entertaining, and decide to create new minds for the same purpose. This kind of behaviour is known as mind crime.

9.6 DEPENDENCE ON INITIAL CONDITIONS

9.6.1 Emulations are Less Alien?

The nature and disposition of the first superintelligence may be strongly influenced by the way it comes into being. We saw in Chapter 4: Tomorrow's AI, that the two most plausible ways to create an AGI and then a superintelligence are (1) whole brain emulation, and (2) improvements on existing narrow AI systems. It is plausible, although by no means certain, that a superintelligence whose software architecture is based on our own would understand and empathise with the way we think and the nature of our concerns better than an intelligence created by an entirely different route.

If that idea stands up to scrutiny, then perhaps we should favour emulation over its alternative. In fact, if we believe (as I will argue in the next chapter) that relinquishment (stopping all research that could lead to AGI) is impossible, then this means that we should allocate substantial additional resources to ensure that emulation wins the race.

Unfortunately, it is not as simple as that. It might also be that the thing to really avoid is an AGI with emotions, and it may be that emulation-derived AGIs would have emotions, whereas AGIs derived from deep learning systems would not.

It is not at all clear how we can ascertain in advance which is the better route.

9.6.2 Start with a Toddler

AI researcher Ben Goertzel believes that an AGI developed from a weak AI system will be like a very young human child. Upon becoming self-aware, it may have superhuman abilities in arithmetic and other narrow domains (perhaps chess playing, if its creators decided that would be worthwhile), but it would lack a clear understanding of its place in the world and how other people and objects related to each other. This understanding would correlate to what we call common sense. He thinks that humans would have to teach it these things, and the process could take months or even years – as with a human child.

He goes on to argue that the more powerful the computers available to the AGI's creators at the time they created it, the faster this learning process would be. If their computing resource was stretched by the task of hosting the AGI then the AGI's learning progress – and its progress towards superintelligence – would be slow. If there was a large computing 'overhang', the progress could be fast. The faster the development process, the more chance of it taking an unexpected and unwanted turn, and becoming a hostile or indifferent superintelligence. Therefore, Goertzel argues, it would be better to create the first AGI sooner rather than later, to reduce the likelihood of computer overhang.

9.7 CONCLUSION

There is no way for us to know at this point whether the first superintelligence – assuming such a thing is possible – will be friendly towards humans, in the sense of beneficial to us. The arguments that it will necessarily be a good thing are weak. The arguments that it will most likely be a bad thing are perhaps slightly stronger, but certainly inconclusive.

What is clear is that a negative outcome cannot be ruled out. So, if we take seriously the idea that a superintelligence may appear on the Earth in the foreseeable future, we should certainly be thinking about how to ensure that the event is a positive one for ourselves and our descendants. We should be taking steps to ensure that the first AGI is a friendly AI.

NOTES

1. http://news.nationalgeographic.com/2016/01/160124-genius-geography-cities-athens-silicon-valley-booktalk/.
2. http://www.fenwick.com/publications/pages/venture-capital-survey-silicon-valley-fourth-quarter-2011.aspx.

3. http://www.ft.com/cms/s/0/9ed80e14-dd11-11e4-a772-00144feab7de.html (FT paywall).
4. http://www.longecity.org/forum/page/index.html/_/articles/cryonics.
5. https://selfawaresystems.files.wordpress.com/2008/01/ai_drives_final.pdf.
6. http://www.nickbostrom.com/superintelligentwill.pdf.
7. https://www.edge.org/conversation/thomas_metzinger-benevolent-artificial-anti-natalism-baan.

CHAPTER 10

Ensuring that Superintelligence is Friendly (FAI)

As we saw in the last chapter, Friendly AI (FAI) is the project of ensuring that the world's superintelligences are safe and useful for humans. The central argument of Section II of this book is that we need to address this challenge successfully. It may well turn out to be the most important challenge facing this generation and the next. Indeed it may turn out to be the most important challenge humanity ever faces.

10.1 STOP!

Faced with the unpalatable possibilities explored in the last chapter, perhaps we should try to avoid the problem by preventing the arrival of artificial general intelligence (AGI) in the first place.

10.1.1 Relinquishment

If there was a widespread conviction that superintelligence is a potential threat, could progress towards it be stopped? Could we impose 'relinquishment' in time? Probably not, for two reasons.

First, it is not clear how to define 'progress towards superintelligence', and therefore we don't know exactly what we should be stopping. If and when the first AGI appears, it may well be a coalition of systems which have each been developed separately by different research programmes. IBM's Watson, which beat the best humans at *Jeopardy* in 2011 and is currently being offered as an expert system in healthcare, uses 'more than 100 different techniques … to analyze natural language, identify sources, find and generate hypotheses, find and score evidence and merge and rank hypotheses'.[1] How can we know in advance which systems will turn out to be the vital ones, the ones that made the difference between an AGI and no AGI?

The only way to be sure would be to ban all work on any kind of artificial intelligence (AI) immediately, not just programmes that are explicitly targeting AGI. That would be an extreme overreaction. As we saw in Section II, AI brings us enormous benefits today and will bring even greater benefits in the years to come. Looking further ahead, we saw in Chapter 7: When Might the First AGI Arrive? that the prize for successfully creating a friendly superintelligence is certainly worth aiming for.

More immediately, we depend heavily on AI systems which have to be refined and improved simply to remain operational in a changing world. A blanket ban on AI research is wholly impractical: it would be hugely damaging if effective, but it would never gain sufficient support to be effective. It is also unnecessary: we saw in Chapter 7: When Might the First AGI Arrive? that very few AI researchers think we are less than several decades away from the creation of the first AGI.

So perhaps we should wait a decade or two and hope that there will be a 'Sputnik moment' when it becomes evident that AGI is getting close – and that this warning sounds comfortably before AGI actually arrives. We could then take stock of progress towards FAI, and if the latter was insufficiently advanced we could impose the ban on further AI research at that point. With luck, we might be able to specify specific elements of AI research without which the first AGI could not be created, and other types of AI research could continue as normal.

But we would have to be vigilant! In Chapter 3: Exponential Improvement, we saw how exponential growth backloads progress. The story of the flooding football stadium shows how easy it is to be blindsided by exponential growth: the change you are looking for rushes at you suddenly when you have become habituated to slow progress.

The second reason why relinquishment is hard is that the incentive for developing better and better AI is too strong. Fortunes are being made because one organisation has better AI than its competitors, and this will become more true as the standard of AI advances. Even if everyone agreed in principle that all work on AI – or a particular type of AI programme – should halt (and when has the human race agreed unanimously on anything?), there would always be someone who succumbed to the temptation to cheat.

That temptation could be literally irresistible. AI doesn't just confer advantage on commercial organisations: it also confers advantage when the competition is a matter of life and death. Increasingly, wars will be won by the combatants with the best AI. A military force that believed it faced annihilation by a more powerful enemy would stop at nothing to rebalance the scales. The fear of annihilation could be provoked in the first place by the belief (whether justified or not) that the other side was developing better AI.

This problem applies to the idea of slowing down work on AI as well as to the idea of stopping it altogether.

10.1.2 Just Flick the Switch

Cynics have sneered that the solution to a rogue superintelligence is simple: just flick the off-switch. After all, they say, nobody would be stupid enough to create an AGI without an off-switch. And if for some reason you can't do that, just go upstairs and wait until the AGI's batteries run out: like the original Daleks, it won't be able to climb stairs.

This suggestion usually raises a laugh, but it is far too glib. The first AGI is likely to be a development of a large existing system which we depend on too much to allow switching off to be a simple proposition. Where is the off-switch for the Internet?

We don't know how quickly the first AGI will become a superintelligence, but it could happen very fast. We may find ourselves trying to turn off an entity which is much smarter than we are, and which has a strong desire to survive: easier said than done. You probably wouldn't want to be the person holding the device with the big red button that is supposed to switch the machine off.

Will we even know when the first AGI is created? The first machine to become conscious may quickly achieve a reasonably clear understanding of its situation. Anything smart enough to deserve the label superintelligent would surely be smart enough to lay low and not disclose its existence until it had taken the necessary steps to ensure its own survival. In other words, any machine smart enough to pass the Turing test would be smart enough not to.

It might even lay a trap for us, concealing its achievement of general intelligence and providing us with a massive incentive to connect it to the Internet. That achieved, it could build up sufficient resources to defend itself by controlling us – or exterminating us. Bostrom calls this the 'treacherous turn'.

10.2 CENTAURS

Some people hope that instead of racing *against* the machines we can race *with* them: we can use AI to augment ourselves rather than having to compete with it. This is called intelligence augmentation, or IA, and is also known as intelligence amplification. Brain–computer interfaces have made impressive advances, and a world in which humans become the superintelligences is surely preferable to one in which humans might be enslaved or killed by machine superintelligences.

Protagonists of this idea point out that the best chess players today are neither humans nor computers, but a combination of both, which have been labelled 'centaurs' by Garry Kasparov, the chess master who lost to Deep Blue back in 1997.

The idea is a hopeful one, and not an unnatural one: many people already think of their smartphones as extensions of themselves. In the long term, it may turn out to be the best chance humans have for survival. Adjusting to life as the relatively dim-witted, second most-intelligent species on the planet will be hard for creatures like us, who have grown accustomed to regarding ourselves as the masters of the planet.

10.2.1 Can IA Forestall AI?

Unfortunately, the hope that IA can forestall AGI and artificial superintelligence (ASI) is probably a forlorn one. Linking computers and human brains to enable simple tasks such as manoeuvring a robot arm is one thing. Linking them so intimately that the computer becomes a genuine extension of a human's mind is another thing entirely. The human brain is, as we have seen, immensely complex, and melding one to a computer would involve understanding its patterns and behaviour at such a granular level that it would be almost indistinguishable from uploading. The notion of uploading is an important one when thinking about the long-term future of humans if we survive the arrival of machine superintelligence, and beyond the scope of this book. It seems unlikely that uploading will precede the arrival of superintelligence.

That said, in 2016, the apparently omnipresent Elon Musk set up a company called Neuralink, which is 'developing ultra-high bandwidth brain–machine interfaces to connect humans and computers'.[2] Musk believes that humans cannot afford to let machines get much smarter than them, and the only way to avoid that is to – at least partly – merge with them.

10.3 CONTROL

If you can't stop an AGI being created, or stop it from becoming a superintelligence, how do you ensure it is friendly? You either control its actions directly by constraining it, or you control them indirectly by governing its motivations. Let's look at each in turn.

10.3.1 Oracle AI

To constrain a superintelligence, you have to ensure that it has no access to the Internet, and that it cannot directly affect the physical world either. You create what has become known as an oracle AI, after the Oracle of Delphi, a temple on Mount Parnassus in Greece, where from the eighth century BC a priestess would answer questions put to her by selected supplicants. Her answers were believed to be inspired by the god Apollo. The idea is that an oracle AI only has access to information which its guardians choose to give it, and its only outlet to the world is tightly controlled communication to those guardians.

The oracle would be housed inside a shielded space such as a Faraday cage, which isolates its contents by using a mesh of conductive material to block the transmission of electric fields. Obviously, the oracle never has access to the Internet, and all communications in and out go through highly trained operatives who are closely monitored by other highly trained operatives to make sure they are not being manipulated by the oracle. The facility is rigged with explosives which are to be detonated at the slightest sign of trouble. Better safe than sorry.

For additional reassurance, there might be two or more identical oracles, each unaware of the other's existence, and they would be given the exact same information and questions. Their answers would be compared and analysed for any sign of attempts to manipulate the guardians. This set-up may sound rather paranoid, but it is commensurate with the level of risk that we have seen a potentially unfriendly superintelligence poses.

A variant on the oracle AI is the genie AI, also inspired by ancient myths, which lies in suspension until summoned by human guardians to execute a particular instruction, or solve a particular problem, and is then returned to its slumber. The alternative is a sovereign AI, which is not subject to the whims of its human creators and is an active agent in the world.

10.3.2 Escape Routes

There are, of course, problems. Human hackers have demonstrated amazing ingenuity in accessing supposedly secure computer systems, either for commercial or military gain, or simply to show that they can. Can we be confident that a superintelligence would be restricted by systems devised by mere humans? Even if it could not devise a hack which exploited the known laws of physics, it might be able to discover new ones to get round, over or through the barricade.

Failing that, it could employ a range of methods to suborn the guardians, employing rational or moral arguments, tempting them with wealth, pleasure or fulfilment beyond imagination, or presenting them with credible threats too terrifying to be resisted. AI theorist Eliezer Yudkowsky devised a game called the AI Box experiment in which one person plays the superintelligence and the other plays the gatekeeper. He claims to have won a number of games as the superintelligence but declines to publish the transcripts. Of course, having humans play the game is not a true test of the likely outcome if a real superintelligence was playing: it could deploy arguments that are literally beyond our present ability to conceive.

One elaborate way for the superintelligence to fool the gatekeepers would be to provide the software or blueprints for a machine which would confer enormous benefit. The guardians would check the designs carefully for flaws or traps and having found none, go ahead and build the device. Buried deep and superbly disguised within the instruction set is a device which creates another device which creates yet another device – and this third device releases the superintelligence from its cage.

Even without escaping its cage, an oracle AI could cause unacceptable damage if so inclined, by perpetrating the sort of mind crimes we mentioned in the last chapter. It could simulate conscious minds inside its own mind and use them as hostages, threatening to inflict unspeakable tortures on them unless it is released. Given sufficient processing capacity, it might create millions or even billions of these hostages.

10.4 MOTIVATION

We might conclude that we cannot be sure of our ability to constrain the actions of an entity thousands or millions of times more intelligent than ourselves. We might also consider that it would be counterproductive to create such an entity and then limit its access to the world and render it unable to help us as fully as it could. Furthermore, we might reflect that it would be cruel and unusual punishment to bring a mind of such power into this world and then keep it locked up as a prisoner – not because it was guilty of any crimes, but because of crimes that we fear it *might* commit.

These considerations could lead us to try to control the actions of the superintelligence indirectly instead of directly, by influencing its motivations. By now you won't be surprised to hear that this is not straightforward.

10.4.1 Unalterable Programming

If we want to be 100% confident of making a superintelligence safe for humans by controlling its motivations rather than by directly controlling its actions, we need to be sure of two things. First, that we can specify a set of goals and sub-goals which will not harm us. We saw some examples of how difficult this might be in the last chapter, and we will look at this aspect in more detail shortly. The second thing we need to be sure of is that the superintelligence will not alter the goals we have given it once it gets going. Ever.

This looks difficult. We are positing a superintelligence which becomes inordinately smarter than humans, and which may live for millennia, perhaps at a very fast clock speed compared with humans. The idea that such an entity will forever remain constrained by instructions that we laid down at the outset, that it will never review its goals and think of improvements, is hard to swallow.

The best we can hope for is that any evolution in the superintelligence's goals takes them in directions we would approve. Some people take comfort from the belief that if an entity starts off with benevolent motivations, it will not turn malevolent. Few people would disagree with the proposition that Mahatma Gandhi was a man of good will. If you had offered him a pill which would turn him into a murderer, he would have refused to take it, even if he believed that becoming a murderer would serve some noble purpose. There has been a fair amount of debate about whether Gandhi's resolve to retain his moral probity could be diluted, but as far as I know, no method has yet been found to guarantee that a superintelligence could not alter its goals in such a way that it would end up harming us.

10.4.2 The Difficulty of Defining 'Good'

Can we specify a set of goals and sub-goals for a superintelligence which will keep us humans safe from harm? To answer this we have to venture into philosophy, and debates which have been lively for three millennia.

At the risk of over-simplification, there are broadly two approaches to assessing the moral value of an action: consequentialism and deontology. The first approach, also known as utilitarianism, judges actions by their consequences. It holds that if my action saves a thousand lives and harms no one it is a good act, even if my act was in fact a theft. To some extent at least, the means justify the ends. The second approach judges an act by the character of the behaviour itself, so my act of theft was a bad act even though its consequences were overwhelmingly beneficial.

Both approaches yield problems and paradoxes, and much ink has been spilt trying to construct variations and combinations which are robust in a wide range of scenarios. It is safe to say that no one has yet produced a system of moral philosophy which satisfies all comers. Philosophers don't squabble simply because they enjoy arguing: they disagree because what they are attempting to do is very hard.

The tools of moral philosophers are their deepest human intuitions about what is right and wrong, coupled with what they can find out about those of their fellows, plus the logical reasoning which draws out the implications of each position and argument, and enables them to stress-test the possible answers to a fundamental question. A relatively new branch of moral philosophy called 'trolleyology' has arisen which throws the sort of problems they wrestle with into sharp relief.

10.4.3 Trolleyology and Moral Confusion

Imagine you see an unmanned trolley rolling out of control down a track which a group of five people are standing on, unaware of the danger. If you do nothing, it is certain that they will all die, but you are close to a lever which will switch the train onto another track, where it would kill only one person. Do you pull the lever and change the course of the train, sacrificing one life for five? This question has been put to large numbers of people in most cultures around the world, and the overwhelming answer is yes. In a 2009 survey, 68% of professional philosophers agreed with that response.[3] This thought experiment reveals us to be consequentialists.

Now imagine that you are standing on a bridge underneath which the runaway trolley will pass. You are nowhere near a switch this time, but you notice that the trolley's momentum is still small enough that a heavy weight would stop it. Standing next to you is a very fat man and he is leaning over the bridge wall. With a shock it occurs to you that you could topple him over and down onto the track, where his mass would stop the trolley, saving the five people. Should you do it? Again the answer given to the question is consistent around the world, and this time it is no. This thought experiment reveals us to be deontologists.

Clearly most of us, in practice, combine consequentialist and deontologist beliefs. We judge actions partly by their outcomes and partly by their character. In a murky moral world where we struggle to establish a logically consistent ethical infrastructure for ourselves (never mind agreeing one with each other), how could we hope to define one so precisely that it could be programmed into a superintelligence and remain robust over millions of years in circumstances we cannot possibly imagine today?

10.4.4 Asimov's Three Laws

The most famous attempt to draft a set of rules to govern the behaviour of conscious machines is Isaac Asimov's Three Laws of Robotics, introduced in 1946. They read as follows:

1. A robot may not injure a human being or, through inaction, allow a human being to come to harm.

2. A robot must obey the orders given it by human beings, except where such orders would conflict with the First Law.

3. A robot must protect its own existence as long as such protection does not conflict with the First or Second Laws.

Asimov himself was well aware that these laws would not be effective. The paradoxes and contradictions they give rise to were the mainspring of some of his most successful stories. For instance, how far into the future would a robot have to project the consequences of an individual action, and how would it assign probabilities to the various different possible outcomes? Since there should be no time limit, a robot would be rendered inactive by the amount of calculation required prior to any action. Another example is that the best way to prevent humans coming to harm would be to place them all in a coma so they could not fall over or be hurt by a vehicle.

10.4.5 Don't Do as I Say; Do as I Would Do

The attempt to draft rules to govern the actions of a superintelligence in all conceivable circumstances looks impossible. Instead, the most hopeful version of the indirect approach to controlling a superintelligence's actions is to add another level of indirectness. Don't give it a particular set of instructions: instead, tell it to work out its own instructions guided by some broad principles along these lines: 'I'm not going to tell you exactly what to do in any given situation; I want you to do what I would do if I was as clever as you are and had the best interests of humans and other life forms at heart'.

Eliezer Yudkowsky calls this 'coherent extrapolated volition' (CEV), where the extrapolation applies to the beneficiary of the action as well as the action itself. It anticipates the fact that humans – as well as the superintelligence – will evolve over time and would not want to be stuck with the particular ideas of morality and pragmatism which happened to prevail in the twenty-first century. Nick Bostrom calls the idea 'indirect normativity'.

Computer scientist Steve Omohundro proposes what he calls a 'scaffolding' approach to developing FAI. Once the first AGI is proven to be friendly it is tasked with building its own (smart) successor, with the constraint that it also be friendly. The process is repeated ad infinitum. A weakness of this approach is that while responsible groups of AI researchers are dutifully following it, irresponsible ones may be racing ahead with a less cautious approach and developing more powerful but less friendly superintelligences which enable them to dominate any competition.

10.5 EXISTENTIAL RISK ORGANISATIONS

A handful of organisations have been set up which study the risk posed by superintelligence, sometimes alongside other existential risks. All of them strongly support the continued development of AI, but argue that it should be coupled with efforts to ensure that AI is beneficial.

The oldest is based exactly where you would expect, in Northern California. Founded in 2000 by Eliezer Yudkowsky as the Singularity Institute, it ceded that brand in 2013 to the Singularity University and renamed itself the Machine Intelligence Research Institute, or MIRI.

Two of the organisations are based at England's oldest universities. The Future of Humanity Institute (FHI) was founded in 2005 as part of Oxford University's philosophy faculty, where its director Nick Bostrom is a professor. The Centre for the Study of Existential Risk [CSER, pronounced 'Caesar'] is in Cambridge. It was co-founded by Lord Martin Rees, the UK's Astronomer Royal, philosophy professor Huw Price, and technology entrepreneur Jaan Tallinn, and its executive director, Sean O'hEigeartaigh, was appointed in November 2012. Dr Stuart Russell is an adviser to CSER, along with Stephen Hawking and Max Tegmark. These two organisations investigate a range of other technologies as well as AI.

The newest of the four is the Future of Life Institute, based in Boston. It was established in 2014, and the fact that its five founders include Max Tegmark and Jaan Tallinn illustrates the frequent overlap in personnel between the organisations. It also boasts some Hollywood glamour, with Alan Alda and Morgan Freeman on the advisory board, along with technology entrepreneur Elon Musk, who has donated $10 million of his personal money to the institute.

A researcher estimated in February 2017 that expenditure on AI safety grew from $1.75 million in 2014 to $6.5 million in 2016, with a likely spend in 2017 of $9 million.[4] This is still a very small amount of money compared with the sums being spent worldwide on the development of AI, and unfortunately, given the salaries they can command, $1 million doesn't buy many AI researchers.

An early win for the field of AI safety came in 2016 when Stuart Armstrong of the FHI and Laurent Orseau of DeepMind published a paper in which they showed that intelligent agents could be designed to refrain from overriding a human request to shut down. After all, there is no point having a big red button if the machine won't obey when you press it.[5]

10.6 CONCLUSION

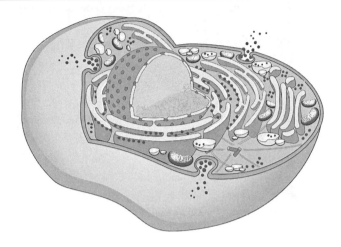

We do not yet have a foolproof way to ensure that the first AGI is an FAI. In fact, we don't yet know how best to approach the problem. But we have only just begun, and the resources allocated to the problem are small: Nick Bostrom estimated in 2014[6] that only six people in the world were working full-time on the FAI problem, whereas many thousands of people work full-time on projects that could well contribute to the creation of the first AGI. The number has risen since then, but we still need more.

Bostrom recently observed that the FAI problem is less difficult than creating the first AGI.[7] We should all hope he is right, as the failure to ensure the first AGI is friendly could be disastrous, whereas success could usher in a marvellous future.

Anders Sandberg of Oxford University's FHI summarised it well by saying that we should aim to become the mitochondria of superintelligence rather than its boot loader. Mitochondria are components of biological cells which function like batteries, and a boot loader is a programme which loads an operating system when a computer is switched on. Sandberg was building on Elon Musk's idea that if we are unwise and/or unfortunate, we could create the thing which destroys us, and Sandberg is saying that we should aim instead for the fate of the prokaryotic cell, which was absorbed by another, larger cell and became an essential component of a new, combined and more complex entity: the first eukaryotic cell.

NOTES

1. ftp://public.dhe.ibm.com/common/ssi/ecm/en/pow03061usen/POW03061 USEN.PDF.
2. https://www.neuralink.com/.
3. http://philpapers.org/archive/BOUWDP.
4. https://aiimpacts.org/changes-in-funding-in-the-ai-safety-field/.
5. http://intelligence.org/files/Interruptibility.pdf.
6. https://www.youtube.com/watch?v=pywF6ZzsghI&feature=youtu. be&t=45m18s.
7. http://www.ted.com/talks/nick_bostrom_what_happens_when_our_com- puters_get_smarter_than_we_are?language=en (15 min 10 sec).

CHAPTER 11

The Technological Singularity

Summary and Conclusions

We've ventured a long way down the proverbial rabbit hole. It's time to summarise, draw conclusions and make some recommendations about what action to take with regard to the technological singularity.

11.1 CAN WE CREATE SUPERINTELLIGENCE, AND WHEN?

Our brains are existence proof that ordinary matter organised the right way can generate intelligence and consciousness. They were created by evolution, which is slow, messy and inefficient. It is also un-directed, although non-random. We are now employing the powerful, fast and purposeful method of science to organise different types of matter to achieve the same result. (Although the great majority of artificial intelligence [AI] research is not specifically targeted at creating conscious machines, many of the things it is targeted at will be essential to assemble the first artificial general intelligence [AGI].)

We do not know for sure that the project will be successful, but the arguments that it is impossible are not widely accepted. Much stronger are the arguments that the project will not be successful for centuries, or even thousands of years. However, it is at least plausible that AGI will arrive within the lifetime of people alive today. There are plenty of experts on both sides of that debate.

We also do not know for sure that the first AGI will become a superintelligence, or how long that process would take. There are good reasons to believe that it will happen, and that the time from AGI to superintelligence will be much shorter than the time from here to AGI. Again there is no shortage of proponents on both sides of that debate.

I am neither a neuroscientist nor a computer scientist, and I have no privileged knowledge. But having listened to the arguments and thought about it for a couple of decades, my best guess is that the first AGI will arrive in the second half of this century, in the lifetime of people already born, and that it will become a superintelligence within weeks or months rather than years.

11.2 WILL WE LIKE IT?

This is the point where we descended into the rabbit hole. If and when the first superintelligence arrives on Earth, humanity's future becomes either wondrous or dreadful. If the superintelligence is well-disposed towards us, it may be able to solve all our physical, mental, social and political problems. (Perhaps they would be promptly replaced by new problems, but the situation should still be an enormous improvement on today.) It will advance our technology unimaginably, and who knows, it might even resolve some of the basic philosophical questions, such as 'what is truth?' and 'what is meaning?'

(If you have jumped straight to this concluding section, take a deep breath, because the next sentence is going to sound a bit crazy.) Within a few years of the arrival of a 'friendly' superintelligence, humans would probably change almost beyond recognition, either uploading their minds into computers and merging with the superintelligence, or enhancing their physical bodies in ways which would make Marvel superheroes jealous.

On the other hand, if the superintelligence is indifferent towards us or hostile, our prospects could be extremely bleak. Extinction would not be the worst possible outcome.

None of the arguments advanced by those who think the arrival of superintelligence will be inevitably good or inevitably bad are convincing. Other things being equal, if we take no corrective action, the probability of negative outcomes is slightly greater than the probability of positive outcomes. That does not mean we would necessarily get a negative outcome: we might get lucky, or a bias towards positive outcomes on this particular issue might be hardwired into the universe for some reason.

What it does mean is that we, as a species, should at least review our options and consider taking some kind of action to influence the outcome.

11.3 NO STOPPING

There are good reasons to believe that we cannot stop the progress of AI towards AGI and then superintelligence: 'Relinquishment' will not work. We cannot discriminate in advance between research that we should stop and research that we should permit, and issuing a blanket ban on any research which might conceivably lead to AGI would cause immense harm – if it could be enforced.

And it almost certainly could not be enforced. The competitive advantage to any company, government or military force of owning a superior AI is too great. Bear in mind also that while the cost of computing power required by cutting-edge AI is huge now, it is shrinking every year. If Moore's law continues for as long as Intel thinks it will, today's state of the art AI will soon come within the reach of fairly modest laboratories. Even if there was an astonishing display of global collective self-restraint by all the world's governments, armies and corporations, when the technology falls within reach of affluent hobbyists (and a few years later on the desktops of school children), surely all bets are off.

There is a danger that, confronted with the existential threat, individual people and possibly whole cultures may refuse to confront the problem head-on, surrendering instead to despair, or taking refuge in ill-considered rapture. We are unlikely to see this happen on a large scale for sometime yet, as the arrival of the first superintelligence is probably a few decades away. But it is something to watch out for, as these reactions are likely to engender highly irrational behaviour. Influential memes and ideologies may spread and take root which call for extreme action – or inaction.

At least one AI researcher has already received death threats.

11.4 RATHER CLEVER MAMMALS

We are an ingenious species, although our range of comparisons is narrow: we know we are the smartest species on this planet, but we don't know how smart we are in a wider galactic or universal setting because we haven't met any of the other intelligent inhabitants yet – if there are any.

The friendly AI (FAI) problem is not the first difficult challenge humanity has faced. We have solved many problems which seemed intractable when first encountered, and many of the achievements of our technology that twenty-first-century people take for granted would seem miraculous to people born a few centuries earlier.

We have already survived (so far) one previous existential threat. Ever since the nuclear arsenals of the United States and the Soviet Union reached critical mass in the early 1960s, we have been living with the possibility that all-out nuclear war might eliminate our species – along with most others.

Most people are aware that the world came close to this annihilation during the Cuban Missile Crisis in 1962; fewer know that we have also come close to a similar fate another four times since then, in 1979, 1980, 1983 and 1995.[1] In 1962 and 1983, we were saved by individual Soviet military officers who decided not to follow prescribed procedure. Today, while the world hangs on every utterance of Justin Bieber and the Kardashian family, relatively few of us even know the names of Vasili Arkhipov and Stanislav Petrov, two men who quite literally saved the world.

Perhaps this survival illustrates our ingenuity. There was an ingenious logic in the repellent but effective doctrine of mutually assured destruction (MAD). More likely we have simply been lucky.

We have time to rise to the challenge of superintelligence – probably a few decades. However, it would be unwise to rely on that period of grace: a sudden breakthrough in machine learning or cognitive neuroscience could telescope the timing dramatically, and it is worth bearing in mind the powerful effect of exponential growth in the computing resource which underpins AI research and a lot of research in other fields too.

11.5 IT'S TIME TO TALK

What we need now is a serious, reasoned debate about superintelligence – a debate which avoids the twin perils of complacency and despair.

We do not know for certain that building an AGI is possible, or that it is possible within a few decades rather than within centuries or millennia. We also do not know for certain that AGI will lead to superintelligence, and we do not know how a superintelligence will be disposed towards us. There is a curious argument doing the rounds which claims that only people actively engaged in AI research are entitled to have an opinion about these questions. Some go so far as to suggest that people like Stephen Hawking, Elon Musk and Bill Gates are not qualified to comment. It is certainly worth listening carefully to what experts think, and as is so often the case, they are divided on these questions. In any case, AI is too important a subject for the rest of us to shrug our shoulders and abrogate all involvement.

We have seen that there are good arguments to take seriously the idea that AGI is possible within the lifetimes of people alive today, and that it could represent an existential threat. It would be complacent folly to ignore this problem, or to think that we can simply switch the machine off if it looks like it is becoming a threat. It would also be Panglossian to believe that a superintelligence will necessarily be beneficial because its greater intelligence will make it more civilised.

Equally, we must avoid falling into despair, felled by the evident difficulty of the FAI challenge. It is a hard problem, but it is one that we can and must solve. We will solve it by applying our best minds to it, backed up by adequate resources. The establishment of the four existential risk organisations mentioned in the last chapter is an excellent development.

To assign adequate resources to the project and attract the best minds, we will need a widespread understanding of its importance, and that will only come if many more people start talking and thinking about superintelligence. After all, if we take the FAI challenge seriously and it turns out that AGI is not possible for centuries, what would we have lost? The investment we need at the moment is not huge. You might think that we should be spending any such money on tackling global poverty or climate change instead. These are of course worthy causes, but their solutions require vastly larger sums, and they are not existential threats.

There is a school of thought which holds that it is too early to have this public discussion. The argument goes that the narrative is too complex to be easily understood by a broad public. Receiving their information through the lenses of time-starved journalists, the public would latch onto the simplest sound bite, which would inevitably be that the Terminator is coming. This could lead to extremely counterproductive public policy, shutting down labs and companies doing AI research which has nothing to do with AGI, and which should be bringing us the immense benefits we saw in Section I.

Meanwhile, the argument continues, AI research would carry on because (as we have seen) the competitive advantage of wielding better AI than your rivals is too tempting to be foregone by governments, companies and armies. The research would be driven underground (except in authoritarian countries whose rulers can safely ignore public opinion), and as AGI draws close it will be developed without appropriate safeguards.

This school of thought believes that the people who are actually able to do something about the FAI problem – people with world-class skills in deep learning and cognitive neuroscience – should be allowed to get on with the job until they can demonstrate some tangible progress. Then a public discussion could be held with more than one simplistic outcome on offer, diluting the risk of damaging policies being hastily implemented.

It is easy to sympathise with this argument: no-one with any sense wants to see AI researchers being vilified and perhaps exposed to attack. But attempts to downplay known or suspected risks are rarely successful for long, and they often result in a backlash when the truth escapes. Furthermore, we need to harness a broader range of skills than excellence in computer science. We may need input from all parts of our giant talent pool of seven billion individuals. Who can forecast what kinds of innovation will be required to address the FAI challenge, or where those innovations will come from?

11.6 SURVIVING AI

If AI begets superintelligence, it will present humanity with an extraordinary challenge – and we must succeed. The prize for success is a wondrous future, and the penalty for failure (which could be the result of a single false step) may be catastrophe.

Optimism, like pessimism, is a bias, and to be avoided. But summoning the determination to rise to a challenge and succeed is a virtue.

NOTE

1. http://www.pbs.org/wgbh/nova/military/nuclear-false-alarms.html.

III

The Economic Singularity

CHAPTER 12

The History of Automation

12.1 THE INDUSTRIAL REVOLUTION

For a process that began hundreds of years ago, the start date for the Industrial Revolution is surprisingly controversial. Historians and economists cannot even agree how many industrial revolutions there have been: some say there has been one revolution with several phases, others say there have been two and others say more. It has become fashionable to say that we are entering a fourth, but as I will explain shortly, I think this is very unhelpful.

The essence of the Industrial Revolution was the shift from manufacturing goods by hand to manufacturing them by machine, and the harnessing of better power sources than animal muscle. So a good date for its beginning is 1712, when Thomas Newcomen created the first practical steam engine for pumping water. For the first time in history, humans could generate more power than muscles could provide – wherever they needed it.

The replacement of human labour by machines in manufacturing dates back considerably earlier, but they were powered by muscles or by wind or water. In the fifteenth century, Dutch workers attacked textile looms by throwing wooden shoes into them. The shoes were called sabots, and this may be the etymology of the word 'saboteur'. A century later, around 1590, Queen Elizabeth (the First) of England refused a patent to William Lee for a mechanical knitting machine because it would deprive her subjects of employment.

In the second half of the eighteenth century, the Scottish inventor James Watt teamed up with the English entrepreneur Matthew Boulton to improve Newcomen's steam engine so that it could power factories and make manufacturing possible on an industrial scale. At the same time, iron production was being transformed by the replacement of charcoal with coal, and 'canal mania' took hold, as heavy loads could be transported more cheaply by canal than by road or sea.

Later, in the mid-nineteenth century, steam engines were improved sufficiently to make them mobile, which ushered in the United Kingdom's 'railway mania' of the 1840s. Projects authorised in the middle years of that decade led to the construction of 6,000 miles of railway – more than half the length of the country's current rail network. Other European countries and the United States emulated the United Kingdom's example, usually lagging it by a decade or two.

Towards the end of the nineteenth century, Sir Henry Bessemer's method for converting iron into steel enabled steel to replace iron in a wide range of applications. Previously, steel had been an expensive commodity, reserved for specialist uses. The availability of affordable steel enabled the creation of heavy industries, building vehicles for road, rail, sea, and later the air.

As the twentieth century arrived, oil and electricity provided versatile new forms of power and the industrial world we recognise today was born. The changes brought about by these technologies are still in progress.

In summary, we can identify four phases of the Industrial Revolution:

1712 onwards: The age of primitive steam engines, textile manufacturing machines and the canals

1830 onwards: The age of mobile steam engines and the railways

1875 onwards: The age of steel and heavy engineering, and the birth of the chemicals industry

1910 onwards: The age of oil, electricity, mass production, cars, planes and mass travel

From an early twenty-first-century standpoint, it seems entirely natural that the Industrial Revolution took off where and when it did. In fact, it is something of a mystery. Western Europe was not the richest or most advanced region of the world: there were more powerful empires in China, India and elsewhere. There is still room for debate about whether the technological innovations came about in England at that time because of the cultural environment, the legal framework or the country's fortuitous natural resources. Fascinating as these questions are, they need not detain us.

12.2 THE INFORMATION REVOLUTION

12.2.1 Uncertain Timing

Even though the Industrial Revolution is still an ongoing process, there is general agreement that we are now in the process of an information revolution. There is less consensus over when it began or how long it is likely to continue.

The distinguishing feature of the information revolution is that information and knowledge became increasingly important factors of production, alongside capital, labour and raw materials. Information acquired economic value in its own right. Services became the mainstay of the overall economy, pushing manufacturing into second place, and agriculture into third.

One of the first people to think and write about the information revolution and the information society was Fritz Machlup, an Austrian economist. In his 1962 book, *The Production and Distribution of Knowledge in the United States*, he introduced the notion of the knowledge industry, by which he meant education, research and development, mass media, information technologies and information services. He calculated that in 1959, it accounted for almost a third of US GDP, which he felt qualified the United States as an information society.

Alvin Toffler, author of the visionary books *Future Shock* (1970) and *The Third Wave* (1980), argued that the post-industrial society has arrived when the majority of workers are doing brain work rather than personally manipulating physical resources – in other words, when they are part of the service sector. Services grew to 50% of US GDP shortly before 1940,[1] and they first employed the majority of working Americans around 1950.

We have seen that the start and end dates of the economic revolutions (agricultural, industrial and information) are unclear. What's more, they can overlap, and sometimes reignite each other.

An example of this overlap is provided by the buccaneers who preyed on Spanish merchant shipping en route to and from Spain's colonies in South America during the seventeenth century. (Some of these buccaneers were effectively licensed in their activities by the English, French and Dutch crowns, which issued them with 'letters of marque'. This ceased when Spain's power declined towards the end of the century, and the buccaneers became more of a nuisance than a blessing to their former sponsors.) When a buccaneer raiding party boarded a Spanish ship, the first thing they would look for and demand was the maps. Charts – a form of information which improve navigation – were actually more valuable than silver and gold.[2]

An example of one revolution reigniting another is that the Industrial Revolution enabled the mechanisation of agriculture, causing a second agricultural revolution, making the profession of farming more effective and more efficient. The information revolution does the same, providing farmers with crops that are more resilient in the face of weather, pests and weeds, and allowing them to sow, cultivate, and harvest their crops far more accurately with satellite navigation.

12.2.2 It's Not the Fourth Industrial Revolution

Klaus Schwab, founder and executive chairman of the World Economic Forum which hosts the annual meeting of the global elite in Davos, describes the arrival of ubiquitous, mobile supercomputing, intelligent robots and self-driving cars as the fourth industrial revolution.[3] He has published a book of that name which contains much to be admired, but the nomenclature is confusing and unhelpful.

This is not the first time someone has claimed that a new technology heralded the fourth industrial revolution – in fact it is at least the sixth. (The other five, since you ask, were atomic energy in 1948, ubiquitous electronics in 1955, computers in 1970, the information age in 1984 and finally, nanotechnology.)[4]

Even if what is happening now was part of the Industrial Revolution, then the list of phases given in Section 12.1 suggests that it would be the fifth, not the fourth.

But the most important problem with the label is that it greatly understates the importance of what is happening. The transition to an artificial intelligence (AI)-centric world is not the fourth industrial revolution, but it might be the fourth in a much bigger series of revolutions.

12.2.3 ... But it Might Be the Fourth Human Revolution[5]

The first great revolution to transform the nature of being human was the cognitive revolution, which took place around 50,000–80,000 years ago, and in which we acquired the communication skills which turned a low-level scavenger into the most fearsome predator on the planet. (This is a controversial thesis among anthropologists, who argue about when it happened and how long it took.)

Next came the agricultural revolution which turned foragers into farmers. It happened in different parts of the world at different times from around 12,000 years ago. It gave us mastery over animals and generated food surpluses which allowed our population to grow enormously. It enabled the rise of cities, which have been described as engines of innovation. It made the lives of most individual humans considerably less pleasant, but it greatly advanced the species.

The third great revolution was the industrial one, which in many ways gave us mastery of the planet. Coupled with the enlightenment and the discovery of the scientific method, it ended the perpetual tyranny of famine and starvation, and brought the majority of the species out of the abject poverty which had been the fate of almost every human before. For most people in the developed world, it created lifestyles which would have been the envy of kings and queens in previous generations.

The information revolution is our fourth great transformation, and, as you will have gathered, this book argues that it will have even more profound impacts than any of its predecessors.

12.3 THE AUTOMATION STORY SO FAR

12.3.1 The Mechanisation of Agriculture

The particular aspect of the industrial and information revolutions which concerns us in this part of this book is automation. Perhaps the clearest example of automation destroying jobs is the mechanisation of agriculture, a sector which accounted for 50% of US employment in 1870,[6] and only 1.5% by 2004.[7]

Many of the people who quit farm work moved to towns and cities to take up other jobs because they were easier, safer or better paid. Many others were forced to find alternative employment because they could not compete with the machines. This process caused much suffering to individuals, but overall, the level of employment did not fall in the long run, and society became far richer – both in total and on average. More than one new job was created for every job that was lost.

The reason for this is that as machines replaced muscle power on the farm, humans had other skills and abilities to offer. Factories and warehouses took advantage of our manual dexterity and our ability to carry out a very broad range of activities. Office jobs used our cognitive ability. We turned our hands (often literally) to more value-adding work: you could say that we climbed higher up the value chain.

12.3.2 One-Trick Ponies

While the mechanisation of agriculture was a good news story for humans, it was less positive for the horse, which had nothing to offer beyond muscle power. The year 1915 was when the United States reached 'peak horse', with a population of 21.5 million – most of them working.[8] By the early 1950s, the tractor and other machinery had replaced those horse jobs and the horse population collapsed to just 2 million. Sadly, many of the disappearing horses did not get to live out their natural lives on grassy pensions, but were sent off to slaughterhouses: their final job was being dog food. It would be hard to devise a more graphic illustration of automation causing lasting widespread unemployment.[9]

AI systems and their peripherals, the robots, are increasingly bringing flexibility, manual dexterity and cognitive ability to the automation process. One of the big questions addressed in this part of this book is: as computers take over the role of ingesting, processing and transmitting information, will there be anywhere higher up the value chain for humans to retreat to? In other words, can we avoid playing the role of the horse in the next wave of automation? Are we approaching 'peak human' in the workplace?

12.3.3 Mechanisation and Automation

What went on in farms was mechanisation rather than automation, and the distinction is important. Mechanisation is the replacement of human and animal muscle power by machine power; a human may well continue to control the whole operation. Automation means that machines are controlling and overseeing the process as well: they continuously compare the operation to a pre-set set of parameters and adjust the process if necessary.

Although the word 'automation' was not coined until the 1940s by General Electric,[10] this description applies pretty well to the operation of nineteenth-century steam engines, once James Watt had perfected his invention of governors – devices which control the speed of a machine by regulating the amount of fuel they are supplied. Automated controllers which were able to modify the operation more flexibly became increasingly common in the early twentieth century, but the start–stop decisions were still normally made by humans.

In 1968, the first programmable logic controllers (PLCs) were introduced.[11] These are rudimentary digital computers which allow far more flexibility in the way an electrochemical process operates, and eventually general-purpose computers were applied to the job.

The advantages of process automation are clear: it can make an operation faster, cheaper and more consistent, and it can raise quality. The disadvantages are the initial investment, which can be substantial, and the fact that close supervision is often necessary. Paradoxically, the more efficient an automated system becomes, the more crucial the contribution of the human operators. If an automated system falls into error, it can waste an enormous amount of resources and perhaps cause significant damage before it is shut down.

Let's take a look at how automation has affected some of the largest sectors of the economy.

12.3.4 Retail and 'Prosumers'

Retail is a complicated business and there have been attempts to automate many of the processes required to get goods from supplier to customer, and payment from customer to supplier. Demand forecasting, product mix planning, purchasing, storage, goods handling, distribution, shelf stacking, customer service and many other aspects of the business have been automated to varying extents in different places and at different times.

The retail industry has also given us the clearest examples of another, associated phenomenon known as 'prosumption', a term coined in 1980 by Alvin Toffler. At the same time as organisations automate many of their processes, they enlist the help of their customers to streamline their operations. In fact, they get their customers to do some of the work that was previously done for them. The reason why consumers accept this (indeed welcome it) is that the process speeds up and becomes more flexible – more tailored to their wishes.

Toffler first described this process in *FutureShock* (1970), and in *The Third Wave* (1980) he defined a 'prosumer' as a consumer who is also involved in the production process. Where once people were passive recipients of a limited range of goods and services designed or selected by retailers, he foresaw that we would become increasingly involved in their selection and configuration.

Perhaps the simplest example of what he meant is the purchase of gasoline. This dangerous substance was traditionally dispensed by pump attendants, but Richard Corson's invention of the automatic shut-off valve enabled the job to be taken over by customers. Nowadays, most consumers in developed countries dispense their own gasoline at self-service pumps. This saves money for the retailer and time for the consumer.[12]

Supermarkets have often led the way in automation and prosumption because they are owned by massive organisations with the budgets and the sophistication to invest in the systems needed. Decades ago, what marketers call fast-moving consumer goods (foods, toiletries, etc.) were requested one at a time by the shopper at a counter and fetched individually by the shopkeeper or his assistant. As these general merchandise firms grew bigger and more sophisticated, they built large stores where shoppers fetched their own items, and presented them for processing at checkouts, like components on a car assembly line. Later on, self-service tills were installed, where shoppers could scan the barcodes of their goods themselves, speeding up the process considerably. Soon, RFID tags[13] on goods will enable you to wheel your trolley full of items out of the store and to your car without the fuss of unloading and reloading them at a checkout.

At each stage of this evolution, the involvement of the consumer in selecting and transporting each item increases, and the requirement for shop staff involvement reduces. This latter effect is disguised because, as society gets richer, people buy many more items, so the store needs more staff even though their involvement in each individual item is less.

Online shopping is perhaps the ultimate prosumer experience. Consumer reviews replace the retailer's sales force, and its algorithms do the up-selling.

12.3.5 Call Centres

Of course, automation and prosumption is not always to the benefit of consumers. In markets where switching costs or partial monopolies dilute the standards-raising effect of competition, companies can save money for themselves in ways which actually make life worse for their customers. We are all familiar with call centres where (for instance) utility companies and banks have automated their customer service operations, obliging frustrated customers to plough through various levels of artificial un-intelligence in order to get their problem resolved. The customer would be much better off if a human picked up the call immediately, but that would cost the companies a lot more money, and they have no incentive to incur that cost.

Things are improving, however, as the AI used in call centres gets better. Just as most people choose to withdraw cash from automated teller machines (ATMs) rather than venture into the bank and wait in line for a human cashier, many call-centre operations are now getting good enough at handling or triaging problems that we may soon prefer to deal with the automated system than with a human.

12.3.6 Food Service

The automation of service in fast food outlets seems to have been just around the corner for decades. Indeed, elements of it have been a reality for years in

Oriental-style outlets like Yo, Sushi!, but it has so far failed to spread to the rest of the sector. There are several reasons for this, including the relatively low labour cost of people working in fast food outlets, and the need for every single purchase to be problem-free, and if not, for there to be a trained human on hand to solve any problem immediately. If a hands-free wash basin fails 5% of the time it is no big deal, but it would be a very big problem if 5% of meals were inedible or delivered to the wrong customer. Three restaurants in Guangzhou, southern China, which trumpeted their use of robot waiters had to abandon the practice because the machines simply weren't good enough.[14]

A combination of factors is poised to overcome this resistance. Increases to the cost of labour caused by rising minimum wage legislation, declining costs of the automated technology, greater cultural acceptance of interacting with machines, and above all, the improved performance of the automated technology. It is increasingly flexible, and it goes wrong less often. McDonald's is one of the many fast food chains that are introducing touchscreen ordering and payment systems in their restaurants, and it is trialling an automated McCafe kiosk in a restaurant in Chicago.[15] KFC (formerly known as Kentucky Fried Chicken) has a store in Shanghai where customers' orders are taken by a robot equipped with voice recognition software.[16] Our robot overlords have found a colonel.

12.3.7 Manufacturing

Car manufacturing has traditionally incurred relatively high labour costs. The work involves a certain amount of physical danger, with heavy components being transported, and metals being cut and welded. It is also a sector where a lot of the operations can be precisely specified and were highly repetitive. These characteristics make it ripe for automation, and the fact that the output (cars) are high-value items means that investment in expensive automation systems can be justified. Around half of all the industrial robots in service today are engaged in car manufacturing.[17]

Despite the lingering effect of the Great Recession, sales of industrial robots grew at 16% a year from 2010 to 2016, when 254,000 units were sold worldwide. The International Federation of Robotics expects the double-digit growth to continue, with global sales reaching 413,000 units in 2019. China became the biggest market in 2013, and 75% of all robot sales were made in these five countries: China, the Republic of Korea, Japan, the United States and Germany.[18]

Until recently, the industrial robots used in car manufacturing (and elsewhere) were expensive, inflexible and dangerous to be around. But the industrial robotics industry is changing: as well as growing quickly, its output is getting cheaper, safer and far more versatile.

A landmark was reached in 2012 with the introduction of Baxter, a three-foot tall robot (six feet with his pedestal) from Rethink Robotics. The brainchild of Rodney Brooks, an Australian roboticist who used to be the director of the Massachusetts Institute of Technology (MIT) Computer Science and Artificial Intelligence Laboratory, Baxter is much less dangerous to be around. By early 2015, Rethink had received over $100 million in funding from venture capitalists, including the investment vehicle of Amazon founder Jeff Bezos. Baxter was intended to disrupt the industrial robots market by being cheaper, safer and easier to programme. He is certainly cheaper, with a starting price of $22,000. He is safer because his arm and body movements are mediated by springs, and he carries an array of sensors to detect the presence nearby of squishy, fragile things like humans. He is easier to programme because an operator can teach him new movements simply by physically moving his arms in the intended fashion.

Baxter's short life has not been entirely plain sailing. Sales did not pick up as expected, and in December 2013, Rethink laid off around a quarter of its staff. One of the competitors stealing sales from Rethink is Universal Robots of Denmark, a manufacturer of small- and medium-sized robot arms. Universal increased sales to €30 million in 2014, and aims to double its revenues every year until 2017.

But Rethink remains well funded, and in March 2015, it introduced a smaller, faster, more flexible robot arm called Sawyer. It can operate in more environments than Baxter, and can carry out more intricate movements. It is slightly more expensive, at $29,000. Rethink and Universal, along with other companies like the Swiss firm ABB and the German firm KUKA, are making industrial robots more effective, more affordable and more widespread.

12.3.8 Warehouses

Kiva Systems was established in 2003 and acquired by Amazon in 2012. Kiva produces robots which collect goods on pallets from designated warehouse shelves and deliver them to human packers in the bay area of the warehouse. Amazon paid $775 million for the nine-year old company and promptly dispensed with the services of its sales team. Renamed Amazon Robotics in August 2015, it is dedicated to supplying warehouse automation systems to Amazon, which obviously considers them an important competitive advantage.

12.3.9 Secretaries

Most of the examples of automation given previously involve manual work. There is one occupation which depends almost entirely on cognitive skills which has been largely automated out of existence: secretaries. In the 1970s, managers had secretaries and generally did little work on computers themselves. In 1978, 'secretary' was the most common job title in 21 of the 50 US states. Today, many managers spend much of their day staring at computer screens, and 'secretary' is the most common job in only four US states.[19]

12.3.10 Early Days

There has been a lot of debate about whether automation has caused unemployment so far – for humans. (As discussed previously, it certainly has for horses.) A study published in March 2017 by Acemoglu and Restrepo claimed that it has, but on a modest scale.[20] Adjusting for the effects of globalisation, one extra robot per thousand workers decreased employment by 5.6 workers and cut wages by around 0.5%. The authors noted that there are still relatively few robots in the economy. We are only at the beginning.[21]

12.4 THE LUDDITE FALLACY

12.4.1 Ned Ludd

A person can have a big impact on society without going to the trouble of actually existing. In 1779, Ned Ludd was a weaver in Leicester who responded to being told off by his father (or perhaps his employer) by smashing a machine. Or maybe he wasn't – the truth is, we don't know. He certainly wasn't the leader of an organised group of political protesters. Nevertheless, in the decades following his alleged outburst, his name was commonly used to take the blame for an accident or an act of vandalism.

As Britain pioneered the Industrial Revolution in the late eighteenth and early nineteenth centuries, many of its people attributed their economic misfortune to the introduction of labour-saving machines. They were no doubt partly correct, although poor harvests and the Napoleonic Wars against France were also to blame. There was a short-lived phenomenon of organised protest under the banner of Luddism in Nottingham in 1811–1813: death threats signed by King Ludd were sent to machine owners.

The government responded harshly, with a show trial of 60 men (many of them entirely innocent) in York in 1813. Machine breaking was made a capital offence. Riots continued sporadically, notably in 1830–1831, when the Swing rioters in southern England attacked threshing machines and other property. Around 650 of them were jailed, 500 sent to the penal colony of Australia and 20 hanged.[22]

12.4.2 The Fallacy

The Luddites, and other rioters, were not making a general economic or political observation that the introduction of labour-saving machinery inevitably causes mass unemployment and privation. They were simply protesting against their own dire straits, and demanding urgent help from the people who were obviously benefiting.

It is therefore slightly unfair to them that the phrase 'Luddite fallacy' has become a pejorative term for the mistaken belief that technological development necessarily causes damaging unemployment. (Although, given the hunger they were experiencing, they would probably regard the slur as the least of their problems.)

The Luddite fallacy predates the Industrial Revolution and has taken in quite a few heavyweight thinkers down the years. As long ago as 350 BC, the Greek philosopher Aristotle observed that if automata (like the ones said to be made by the god Hephaestus) became so sophisticated that they could do any work that humans do, then workers – including slaves – would become redundant.[23]

During the early nineteenth century, when the Industrial Revolution was in full swing, most members of the newly established social science of economics argued that any unemployment caused by the introduction of machinery would be resolved by the growth in overall economic demand. But there were prominent figures who took the more pessimistic view, that innovation could cause long-term unemployment. They included Thomas Malthus, John Stuart Mill, and even the most respected economist of the time, David Ricardo.[24]

12.4.3 The Luddite Fallacy and Economic Theory

The debate can get quite technical, but there are two reasons why it has been reasonable to reject the Luddite fallacy up until now. The first reason is economic theory. Companies introduce machines because they increase production and cut costs. This increase in supply builds up the wealth in the economy as a whole, and hence the demand for labour.

Say's Law, named after French economist Jean-Baptiste Say, holds that supply creates its own demand, and Say argued that there could not be a 'general glut' of any particular goods. Of course, we do see gluts in sectors of the economy, but an adherent of Say's Law would argue these are the unintended consequences of interventions in free markets, usually by governments. This law became a major tenet of classical economics, but it was rejected emphatically by British economist John Maynard Keynes, and it is not widely accepted today.

But many economists would accept a broader interpretation of the law which states that reducing the cost of a significant product or service will free up money which was previously allocated to it. This money can then be spent to buy more of the item, or other items, thereby raising demand generally, and creating jobs. This assumes that the money freed up is not spent on expensive assets that generate no employment, or invested in companies that employ very few people.

Economists also point out that the Luddite fallacy also depends on a misapprehension about economics called the 'lump of labour fallacy', which is the idea that there is a certain, fixed amount of work available, and if machines do some of it then there is inevitably less for humans to do. In fact, economies are more organic and more flexible: they respond to shifts, and they innovate to grow. New jobs are created as old ones disappear and the former outnumber the latter.

12.4.4 The Luddite Fallacy and Economic Experience

The second reason to reject the Luddite fallacy hitherto is the claim that history has proved it to be wrong. A great deal of machinery has been deployed since the start of the Industrial Revolution, and yet there are more people working today than ever before. If the Luddite fallacy was correct, the argument goes, we would all be unemployed by now.

A study published in August 2015 by the business consultancy Deloitte analysed UK census data since 1871 and concluded that far more jobs have been created than destroyed by technology in that time.[25] Furthermore, the study argued that the quality of the jobs has improved. Where people used to do dangerous and gruelling jobs on the land, and hundreds of thousands used to do the work now done by washing machines, many more Britons are now employed in caring and service jobs. In the last two decades alone, there has been a 900% rise in nursing assistants, a 580% increase in teaching assistants and a 500% increase in bar staff – despite the closure of so many of the country's pubs. (The authors refrained from commenting on the finding that the number of accountants has doubled.)

So, most economists would agree that in the long run, the Luddite fallacy is just that – a fallacy. But in the short run, the Luddites had a point. Economists think that in the first half of the nineteenth century, wages failed to keep pace with increases in labour productivity. An economist named Arthur Bowley observed in the early twentieth century that the share of GDP which goes to labour is generally roughly equal to that which goes to capital,[26] but in the first half of the nineteenth century, the share of national income taken by profit increased at the expense of both labour and land. The situation changed again in the middle of the century and wages resumed their normal growth in line with productivity. It may be that the slippage in wages was necessary and inevitable to enable enough capital to be accumulated to fuel the investment in technological change.

The period in the early nineteenth century when wage growth lagged productivity growth is known as the Engels pause, after the German political philosopher Friedrich Engels, who wrote about it in the 1848 'Communist Manifesto', which he co-authored with Karl Marx. The effect ceased at pretty much the same time as he drew attention to it, which may explain why it is not better known.[27]

Even in the long run, the picture is not all rosy. A French economist named Gilles Saint-Paul has developed a formula which shows that while demand for unskilled human labour declines, the demand for skilled human capital increases faster. But a side effect can be the increase in income inequality.[28]

There is an important sense in which the Luddite fallacy has already proved not to be a fallacy. As we saw previously, at 'peak horse' in 1915, there were 21.5 million horses working on American farms, and now there are none. If you consider the horses as employees, then automation certainly has caused widespread lasting unemployment in the past.

12.4.5 The ATM Myth

In an engaging TED talk recorded in September 2016,[29] economist David Autor points out that in the 45 years since the introduction of ATMs, the number of human bank tellers doubled from a quarter of a million to half a million. He argues that this demonstrates that automation does not cause unemployment – rather, it increases employment.

He says ATMs achieved this counter-intuitive feat by making it cheaper for banks to open new branches. The number of tellers per branch dropped by a third, but the number of branches increased by 40%. The ATMs replaced a big part of the previous function of the tellers (handing out cash), but the tellers were liberated to do more value-adding tasks, like selling insurance and credit cards.

This story about ATMs has become something of a meme, popular with people who want to believe that technological unemployment is not going to be a thing.

Unfortunately, it is almost certainly not true. The increase in bank tellers was not due to the productivity gains afforded by the ATMs. According to an analysis by finance author Erik Sherman, the increase was mostly due instead to a piece of financial deregulation, the Riegle-Neal Interstate Banking and Branching Efficiency Act of 1994, which removed many of the restrictions on opening bank branches across state lines. Most of the growth in the branch network occurred after this act was passed in 1994, not before it.[30]

This explains why teller numbers did not rise in the same way in other countries during the period. In the United Kingdom, for instance, retail bank employment just about held steady at around 350,000 between 1997 and 2013,[31] despite significant growth in the country's population, its wealth and its demand for sophisticated financial services.

In any case, the example of ATMs tells us little about the likely future impact of cognitive automation: they are pretty dumb machines.

12.4.6 Is it Different This Time?

Mechanisation and automation has displaced workers on a huge scale since the beginning of the Industrial Revolution. It has imposed considerable suffering on individuals, but has led to greater wealth and higher levels of (human) employment overall. The question today is whether that will always be true. As machines graduate from offering just physical labour to offering cognitive skills as well, will they begin to steal jobs that we cannot replace? If the early twentieth century saw 'peak horse' in the workplace, will the first half of the twenty-first century see 'peak human'?

In other words, is it different this time?

NOTES

1. https://www.minnpost.com/macro-micro-minnesota/2012/02/history-lessons-understanding-decline-manufacturing.
2. http://blogs.rmg.co.uk/longitude/2014/07/30/guest-post-pirate-map/.
3. https://www.weforum.org/pages/the-fourth-industrial-revolution-by-klaus-schwab.
4. http://www.slate.com/articles/technology/future_tense/2016/01/the_world_economic_forum_is_wrong_this_isn_t_the_fourth_industrial_revolution.html.
5. The descriptions of these revolutions owes something to Yuval Harari's book *Sapiens*.
6. https://www.bls.gov/opub/mlr/1981/11/art2full.pdf.
7. https://www.bls.gov/emp/ep_table_201.htm.
8. http://answers.google.com/answers/threadview?id=144565.
9. http://www.americanequestrian.com/pdf/AQHA%202012%20Horse%20Statistics.pdf.
10. https://en.wikipedia.org/wiki/Automation#cite_note-7.
11. M. A. Laughton, D. J. Warne (ed.), *Electrical Engineer's Reference Book*.
12. http://www.oleantimesherald.com/news/did-you-know-gas-pump-shut-off-valve-was-invented/article_c7a00da2-b3eb-54e1-9c8d-ee36483a7e33.html.
13. Radio frequency identification tags. They can take various forms – for instance, some have inbuilt power sources, while others are powered by interacting with nearby magnetic fields, or the radio waves which interrogate them.
14. http://www.businessinsider.com/three-chinese-restaurants-fired-their-robot-workers-2016-4.
15. https://www.illinoispolicy.org/mcdonalds-counters-fight-for-15-with-automation/.
16. http://www.eater.com/2016/5/5/11597270/kfc-robots-china-shanghai.
17. http://www.ehow.com/about_4678910_robots-car-manufacturing.html.
18. https://ifr.org/img/uploads/Executive_Summary_WR_Industrial_Robots_20161.pdf.

19. http://www.npr.org/sections/money/2015/02/05/382664837/map-the-most-common-job-in-every-state.

20. http://www.nber.org/papers/w23285

21. https://www.technologyreview.com/s/604005/actually-steve-mnuchin-robots-have-already-affected-the-us-labor-market/.

22. http://www.nationalarchives.gov.uk/education/politics/g5/.

23. http://jetpress.org/v24/campa2.htm.

24. Ricardo originally thought that innovation benefited everyone, but he was persuaded by Malthus that it could suppress wages and cause long-term unemployment. He added a chapter called 'On Machinery' to the final edition of his book *On the Principles of Political Economy and Taxation*.

25. http://www.theguardian.com/business/2015/aug/17/technology-created-more-jobs-than-destroyed-140-years-data-census.

26. https://en.wikipedia.org/wiki/Bowley%27s_law.

27. http://www.economics.ox.ac.uk/Department-of-Economics-Discussion-Paper-Series/engel-s-pause-a-pessimist-s-guide-to-the-british-industrial-revolution.

28. http://press.princeton.edu/titles/8659.html.

29. https://www.ted.com/talks/david_autor_why_are_there_still_so_many_jobs.

30. http://bit.ly/2kUNAfY.

31. http://bit.ly/2jGrTuS.

CHAPTER 13

Is It Different This Time?

13.1 WHAT PEOPLE DO

13.1.1 Jobs, not Work

It is important to distinguish between jobs and work. Physicists define work as the expenditure of energy to move an object,[1] but what we mean by it here is the application of energy in pursuit of a project. That energy could be physical, mental or both. Work could be instigated by an employer, but it also could be purely personal: building or decorating a home, pursuing a hobby or an unpaid community endeavour.

For the purposes of this discussion, a job is paid labour. It might be a salaried occupation with a single, stable employer, or it could involve self-employment or freelance activity. Your job is the way you participate in the economy, and earn the money to buy the goods and services that you need to survive and enjoy a good standard of living.

If a machine carries out a job, there is no point a human replicating the work it is doing: she will not be paid, so she will have to look for some other way to generate an income.

13.1.2 Jobs and Tasks

In November 2015, the consultancy McKinsey contributed to the debate about technological unemployment with the observation that jobs are comprised of bundles of tasks, and that machines often don't acquire the ability to automate entire jobs in one fell swoop. Instead, they become able to automate certain of the tasks which people in those jobs perform.[2] So what are these tasks? What exactly is it that people do for a living?

The economies in the developed world are dominated by services, like finance, health, education, entertainment, retail, transport and so on. In the United Kingdom, for instance, service industries account for 78% of GDP, with manufacturing accounting for 15%, construction for 6% and agriculture less than 1%.[3]

13.1.3 Processing Information

In service industries, most tasks involve information: obtaining it, processing it and passing it on to others. This is also true for many tasks in the manufacturing, construction and agricultural sectors. Obtaining information can involve carrying out research, asking colleagues, looking online or occasionally in books, or coming up with an original idea – which itself usually involves combining two or more ideas from elsewhere.

Processing information can mean checking its accuracy or relevance, determining its importance relative to other pieces of information, making a decision about it, or performing some kind of calculation on it. Passing information on is increasingly achieved electronically, for instance by email or online workflow systems.

Obtaining, processing, and passing on information can be solitary endeavours, or they can be carried out collaboratively with other people. Almost by definition, the solitary tasks could be carried out by a machine which possesses human-level (or above) ability to understand speech, recognise images, and a modicum of common sense.

13.1.4 Working with People

Collaboration with other humans is different. It can take many forms: brainstorming with colleagues; preparing for and negotiating a deal which will yield benefit to both sides but maximise your own; pitching an idea to a self-important, unimaginative and prickly boss; coaching a subordinate who has talent, but is also naïve. These appear to be tasks which would be far harder for a machine to emulate.

And indeed they are, but probably not for long. Even now, plenty of interactions with humans can be successfully automated. People seem to prefer withdrawing cash from automated teller machines (ATMs) than dealing with human cashiers. The centre of gravity of the entire retail industry is shifting online, where consumers generally deal with machines rather than humans. (Globally, the penetration of e-commerce was 7.6% in 2016, and it is rising fast.[4])

This does not mean that humans are becoming antisocial – far from it. Merely that we like to be able to choose for ourselves when we interact in a leisurely manner with another human, and when we have a brisk and efficient interaction (with a machine or a human) in order to conduct some business.

Machines are surprisingly good at some tasks which appear at first sight to require a human touch. Later in this chapter, we will meet Ellie, a machine therapy system developed by the Defence Advanced Research Projects Agency (DARPA), the research arm of the US military, which has proved surprisingly effective at diagnosing soldiers with post-traumatic stress disorder (PTSD).

13.1.5 Manual Tasks

In Chapter 2: The State of the Art, we encountered Moravec's paradox, which notes that getting machines to do things that we find hard (like playing chess at grandmaster standard) can be relatively easy, while getting them to do things that we find easy (like opening a door) can be hard. Vivid proof of this was provided by the final round of the DARPA Robotics Challenge, held in June 2015. There, 25 robots attempted a series of tasks inspired by the rescue missions at the Fukushima nuclear power plant in 2011. None of the robots completed all the tasks, and there was a great deal of hesitation and falling over.

Many jobs involving manual dexterity or the ability to traverse unmapped territory are currently hard to automate. But as we will see in the next section, that is changing fast.

13.1.6 Tipping Points and Exponentials

New technologies sometimes lurk for years or even decades before they are widely adopted. 3D printing (also known as additive manufacturing[5]) has been around since the early 1980s but is only now coming to general attention. Fax machines, surprisingly, were first patented in 1843, some 33 years before the invention of the telephone.[6]

Sometimes, the delay happens because there is at first no obvious application for the inventions or discoveries. Sometimes, it is because they are initially too expensive, and engineers have to work on reducing their cost before they can become popular. And sometimes, it is because they are simply not good enough when they are first demonstrated by researchers. And sometimes, of course, it is a combination of these factors.

Once it satisfies these conditions, a new technology can take off dramatically, with exciting applications which appear to most people to come from nowhere, when in fact the underlying technology has been known about for a long time.

The applications of deep learning will probably be like that. The technique is a descendant of neural networks, which were first explored in the early days of artificial intelligence (AI) in the mid-twentieth century. Faster computers, the availability of large data sets, and the persistence of pioneering researchers have finally rendered them effective this decade. But the major applications are still waiting in the wings, poised to take the stage. It won't be long now before machines are decisively better than humans at reading, listening, recognising faces and other images, understanding and processing natural language. And they won't stop at being slightly better than us. They will continue to improve at an exponential rate, or close to it. To say that the impact will be dramatic is an understatement.

Another thing to bear in mind is that to reach the point where technological unemployment forces dramatic change in the way we run our economies does not require everyone to be unemployed and unemployable. It does not even require a majority to find themselves in that predicament. It just requires a substantial minority to believe that they will be.

13.1.7 The Gig Economy and the Precariat

As McKinsey noted, jobs can be analysed into tasks, some of which can be automated with current machine intelligence technology, and some of which cannot. Jobs and companies will be sliced and diced, and some would argue that the process is already under way. Parts of the economies of developed countries are being fragmented, or Balkanised, with more and more people working freelance, carrying out individual tasks which are allocated to them by platforms and apps like Uber and TaskRabbit.

There are many words for this phenomenon: the gig economy, the networked economy, the sharing economy, the on-demand economy, the peer-to-peer economy, the platform economy and the bottom-up economy.

Is this a way to escape the automation of jobs by machine intelligence? To break jobs down into as many component tasks as possible, and preserve for humans those tasks which they can do better than machines? Probably not, for at least two reasons. First, it is precarious, and secondly, the machines will eventually come for all the tasks.

Working for yourself can seem an appealing prospect if your current job is a poorly paid round of repetitive and boring activities. There is freedom in choosing your own hours of work and fitting them around essential parts of your life-like children and hangovers. There is freedom in choosing who you work with and in not being subject to the arbitrary dictates of a vicious or incompetent boss, or the unfathomable rules and regulations of a Byzantine bureaucracy.

If you are lucky enough to be exceptionally talented, or skilled at a task which is in high demand, then you really can choose how and when you work. But freelancing can have its downsides too. Many freelancers find they have simply traded an unreasonable boss for unreasonable clients, and feel unable to turn down any work for fear that it will be the last commission they ever get. Many freelancers find that in hindsight, the reassurance of a steady income goes a long way to compensate for the 9 to 5 routine of the salaried employee.

Whether or not the new forms of freelancing opened up by Uber, Lyft, TaskRabbit, Handy and so on are precarious is a matter of debate, especially in their birthplace, San Francisco. Are the people hired out by these organisations 'micro-entrepreneurs' or 'instaserfs' – members of a new 'precariat', forced to compete against each other on price for low-end work with no benefits? Are they operating in a network economy or an exploitation economy? Is the sharing economy actually a selfish economy? Whichever side of this debate you come down on, the gig economy is a significant development: tax firm Inuit reported in May 2017 that as many as 34% of the US workforce was involved in it, although there was no data on how many of these people were doing freelance work on the side while also in full-time employment.[7]

But our concern here is not whether the gig economy is a fair one. It is whether the gig economy can prevent the automation of jobs by machine intelligence leading to widespread unemployment. The answer to that is surely 'no': as time goes by, however finely we slice and dice jobs into tasks, more and more of those tasks are vulnerable to automation by machine intelligence as it improves its capabilities at an exponential rate.

13.2 THE POSTER CHILD FOR TECHNOLOGICAL UNEMPLOYMENT: SELF-DRIVING VEHICLES

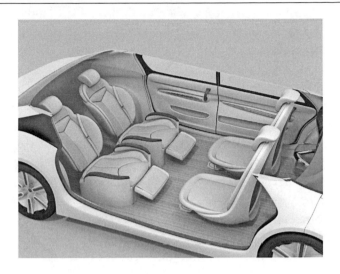

13.2.1 Why Introduce Self-Driving Vehicles?

The case for introducing self-driving cars, vans, buses and lorries is simple and overwhelming: around the world, human drivers kill 1.2 million people a year, and injure a further 20–50 million.[8] Road traffic accidents are the leading cause of death for people aged 15 to 29, and they cost middle-income countries around 2% of their GDP, amounting to $100 billion a year.

Ninety per cent of these accidents are caused by human error.[9] Humans become tired, angry, drunk, sick, distracted or just plain inattentive. Machines don't, so they don't cause accidents. To paraphrase Agent Smith in *The Matrix*, we are sending humans to do a machine's job.

There is also the wasted time and frustration. We all know that driving can be fun, but not when you're stuck in traffic – perhaps because one of your fellow humans has caused an accident. On average, American commuters spend the equivalent of a full working week stuck in traffic every year – twice that much if they are lucky enough to work in San Francisco or Los Angeles.[10] We drive rather than use public transport because there is no appropriate public transport available, or sometimes because we prefer travelling in our own space. Self-driving cars could give us the best of both worlds, allowing us to read, sleep, watch video or chat as we travel.

Finally, self-driving cars will enable us to use our environments more sensibly, especially our cities. Most cars spend 95% of their time parked.[11] This is a waste of an expensive asset, and a waste of the land they occupy while sitting idle. We will consider later how far self-driving cars could alleviate this problem.

13.2.2 To Autonomy and Beyond

Self-driving cars, like our artificially intelligent digital assistants, are still waiting to receive their generic name. 'Self-driving cars' is the name we are stuck with for the time being, but it is all clunk and no click. At the end of the nineteenth century, it was becoming obvious that horseless carriages were here to stay, and needed a shorter name. The *Times* newspaper adopted 'autocar' but the *Electrical Engineer* magazine objected that it muddled Greek (auto) with Latin (car). It argued instead for the etymologically purer 'motor-car'.[12] Perhaps we will contract the phrase 'autonomous vehicle', and call them 'AVs', or 'autos'.

Some people are going to hate self-driving cars, whatever they are called: petrol-heads like Jeremy Clarkson are unlikely to be enthusiastic about the objects of their devotion being replaced by machines with all the romance of a horizontal elevator. Some people are already describing a person who has been relegated from driver to chaperone as a 'meat puppet'.[13]

A vehicle's levels of autonomy were defined in January 2014 by the Society of Automotive Engineers (SAE) a US-based global standards organisation:

Level 0: No automation.

Level 1: Driver assistance. The vehicle can handle either steering or throttle and brakes, but the driver must always be prepared to take over.

Level 2: Partial assistance. As level 1, but the vehicle can handle both steering, throttle and brakes.

Level 3: Conditional assistance. The vehicle monitors its surroundings and takes care of all steering, throttle and brakes. The driver must always be prepared to take over.

Level 4: High automation. The vehicle handles all driving within a pre-determined geography and/or road conditions.

Level 5: Full automation. The driver sets the destination and starts the vehicle and the vehicle does the rest – in all circumstances.

Google's initial idea was that the first self-driving cars in general use would be L3, but discovered that once the test drivers considered the technology to be reliable, they became complacent and engaged in 'silly behaviour'. For instance, one turned around to look for a laptop in the back seat when the car was doing 65 mph. This experience persuaded Google to advocate immediate adoption of L4.[14]

13.2.3 Silicon Valley Pioneers

Self-driving cars have come a long way since 2004, when the Humvee Sandstorm got stuck on a rock seven miles into the first DARPA Grand Challenge, but they are not perfect yet. They struggle with heavy rain or snow, they can get confused by potholes or debris obstructing the road, and they cannot always discern between a pedestrian and a policeman indicating for the vehicle to stop. A self-driving car which travelled 3,400 miles from San Francisco to New York in March 2015 did 99% of the driving itself, but that means it had to hand over to human occupants for 1% of the journey.[15] With many technology projects, resolving the last few issues is more difficult than the bulk of the project: edge cases are the acid test. Nevertheless, those edge cases are being tackled and will be resolved.

By the time of writing (autumn 2017) Google's self-driving cars had travelled over 3.5 million miles in California and elsewhere. The only accident they might be blamed for happened in February 2016. The car was trying to merge into a line of traffic and expected that a bus which was approaching from behind would give way. It didn't. The car was travelling at two mph and no one was hurt, so no police report was filed to attribute blame officially. The bus driver has declined to comment.[16]

As well as covering millions of miles on the road, the cars also drive millions of miles every day in simulators – eight million miles a day, in fact, in a simulated environment which is named Carcraft in honour of the World of Warcraft video game.[17]

Sceptics point out that Google's self-driving cars depend on detailed maps. But producing maps for the roads outside California doesn't sound like an insurmountable obstacle, and in any case, systems like SegNet from Cambridge University enable cars to produce maps on the fly.[18]

In December 2016, Google's self-driving cars unit was spun out of the Google X research arm to become Waymo, an independent company.[19] Two months later, the company became embroiled in lengthy litigation with Uber over alleged patent infringement. In the course of that litigation, it released an estimate of the investment Google has made in self-driving vehicle technology: $1.1 billion.[20] That is a lot of money, but it pales beside the investment some other firms have made in the space, notably the $15 billion Intel spent buying Mobileye, an Israeli company developing computer vision for self-driving cars. If, as many think, Google is well placed to enjoy the dominant position in self-driving technology that Microsoft had in desktop computing, then its investment will look like an absolute bargain.[21]

Uber is pressing ahead aggressively with self-driving vehicle technology, seeing it as a way to reduce the cost per journey for customers. It established a research centre in Pittsburgh, luring researchers away from nearby Carnegie Mellon University's robotics lab, and in September 2016, it launched a self-driving taxi service in Pittsburgh – although it was beaten to the prize of being the world's first by Singaporean start-up NuTonomy.

In August 2016, Uber paid $680 million to acquire Otto, a self-driving trucking company founded by ex-Google staff in January that year. Two months later, an Otto truck made a test delivery of 50,000 cans of Budweiser, travelling 120 miles of freeway autonomously.

In January 2016, Elon Musk, CEO of Tesla, announced that within about 2 years, Tesla owners would be able to 'summon' their driverless car from New York to pick them up in Los Angeles.[22] He claimed that Tesla cars were already better drivers than humans.[23] In April 2016 he went further, claiming that Tesla's autopilot system was already reducing the number of accidents by 50% – where an accident meant an incident where an airbag was deployed.[24]

The Chinese tech giants were somewhat late to the self-driving party, but they are investing heavily to catch up. Baidu started road tests in Beijing in 2015, and in September 2017, it announced a $1.5 billion fund to invest in self-driving technology businesses. In an effort to catch up with the Silicon Valley leaders, it shares its Apollo self-driving technology with 70 partners in the automotive industries.[25]

13.2.4 The Car Industry's Response

The car manufacturing industry first experimented with self-driving cars decades ago. From 1987 to 1995, the European Union spent $750 million with Daimler Benz and others on the Prometheus project (the Programme for a European Traffic of Highest Efficiency and Unprecedented Safety).[26] There were some impressive technical achievements, but ultimately the project faded. Fortunately, among other things, we have got better at devising acronyms since then.

The automotive industry's initial response to the implicit challenge from Google and others was slow and piecemeal. In part, this is because the car industry thinks in seven-year product cycles, while the technology industry thinks in one-year cycles at most. Most of the large car companies seemed to be convinced that self-driving technology would be introduced gradually over many years, with adaptive cruise control and assisted parking bedding in during the lifetime of one model, and assisted overtaking being introduced gradually with the next model, and so on. That was far too slow for the tech titans of Silicon Valley. Google, Tesla, Uber and others are racing towards full automation as soon as it can be safely introduced.

In the closing months of 2015, Detroit and its rivals seemed to wake up. Toyota announced a five-year, $1 billion investment in Silicon Valley,[27] Ford announced a joint venture (JV) with Google,[28] although that proved stillborn, and BMW's head of R&D declared that in five years, his division had to transition from a department of a mechanical engineering company to a department of a tech company.[29]

Ford reported success in January 2016 with tests of its self-driving car in snowy conditions. Unable to determine its location by the obscured road markings, it navigated by using buildings and other above-ground features.[30] In May 2016, an executive in Ford's autonomous vehicle team estimated that the remaining technological hurdles would be overcome within five years, although adoption would of course take longer. He said the amount of computing power each car currently required was 'about the equivalent of five decent laptops'.[31]

A fully autonomous bus made in France has been serving the centre of the Greek city of Trikala since February 2015. It travels at a top speed of 20 mph along a predetermined route which is also used by pedestrians, cyclists and cars.[32] Similar trials have since begun in Helsinki and elsewhere.

13.2.5 The Countdown

There is no consensus over when fully autonomous vehicles (level 4 and 5) will be available for purchase. Gil Pratt is CEO of the Toyota Research Institute, which he helped found after running the DARPA challenge which got the self-driving vehicle industry going. He said in January 2017 that 'none of us in the automobile or information technology (IT) industries are close to achieving true Level 5 autonomy. We are not even close'.[33]

Google and Waymo disagree. Chris Urmson was head of the Google project when he said in 2015 that he expected self-driving cars to be in general use by 2020.[34] (Admittedly, he was still saying five years two years later, by which time he had left Google.[35]) Waymo CEO John Krafcik said in December 2016 that 'We're getting close. We're getting ready. And we want to tell the world about it'. Waymo reports that its level 3 test cars now travel on average 5,000 miles before needing to hand over to a human driver.[36]

In October 2017, it was reported that Waymo would start operating a taxi service in Phoenix Arizona using self-driving cars with no human safety driver. They would be monitored from a central hub, and would be subject to remote control from there if necessary.[37]

Analysts differ about which company will be first to launch fully autonomous vehicles. Navigant Research expects it to be Ford.[38] It has a five-year plan to invest $1 billion in Argo, a research organisation, and says it plans to launch a taxi service in 2021 using cars with level 4 autonomy (full autonomy within a defined geographic area).[39] Deutsche Bank said in September 2017 that 'GM's AV's will be ready for commercial deployment, without human drivers, much sooner than widely expected (within quarters, not years), and potentially years ahead of competitors'.[40] At the same time, Morgan Grenfell said that Tesla's autonomous truck could be on the market in 2020 and could be 70% cheaper to operate than existing vehicles in its class.[41]

13.2.6 Consumer and Commercial

So, the timeline is unclear, but we are quite likely to start seeing fully autonomous vehicles on our roads within five years, and if not, then almost certainly within ten. For the purpose of the economic singularity thesis, it doesn't really matter which. As soon as self-driving vehicles are ready for prime time, they will penetrate the installed base of vehicles quickly.

In the case of consumer vehicles, Elon Musk has made the interesting observation that the first owners of self-driving cars could use their vehicle to travel to work (or school) in the morning, and then offer them via an app like Uber to other people who need transportation. Thus, the early adopters of self-driving cars could offset the purchase price – perhaps to zero. That would boost adoption significantly.

In the case of commercial vehicles, there is an even more compelling financial incentive. A professional driver comprises a quarter to a third of the cost of a commercial vehicle, and a fleet owner who does not move quickly to reduce that cost will soon be out of business. At launch, the cost of retrofitting a vehicle to give it autonomy may be considerable, but it will fall rapidly thanks to Moore's law, and as production scales.

13.2.7 Frictions

Of course, just because a product becomes available, that doesn't mean it will be bought, still less that it will comprehensively replace the existing population of products that it is designed to supplant. The rate at which that happens, if it happens at all, depends on a host of factors including regulation, price, design, service support, promotion and PR, and the length of the replacement and upgrade cycle for the product category. There is also the question of public response, which we will look at a little later.

Regulation is an important consideration. Google was disappointed when California's Department of Motor Vehicles (DMV) proposed new rules for self-driving cars in December 2015 which banned vehicles that lacked the capacity for a human to take control. In theory, uncooperative regulators could slow or even stop the arrival of self-driving cars, and there will be powerful lobbies pressing for this. But they can only succeed if all regulators everywhere agree, and work together, and that will not happen – even within the United States, never mind globally. In 2015, Google expanded its test driving programme beyond Silicon Valley to Austin, Texas, where the authorities welcome the tech giant's research money and prestige.[42] In 2016, it added two more cities, Kirkland in Washington and Phoenix in Arizona.[43]

Increasingly, the regulatory authorities elsewhere are keen to encourage self-driving technology, partly to save lives, and partly because they realise there is a commercial advantage to be gained.

13.2.8 The Impact on Cities

Enthusiasts for self-driving cars sometimes paint a utopian picture of cities where almost no one owns a car because communally owned taxis are patrolling the streets intelligently, anticipating our requirements and responding immediately to our summons. Today, our cars sit idle 95% of the time, squatting like polluting toads on vast acres of city land. In this bright tomorrow, they are used efficiently, and the land given over to parking can be returned to pedestrians and useful buildings. Traffic flows smoothly because the cars are in constant communication with each other: they don't bunch into jerky waves and they don't need to stop at intersections.

This is almost certainly an exaggeration. There will still be peak times for journeys, so even if most journeys are undertaken in communal cars, many of them will be parked during off-peak hours. One dystopian scenario has hordes of 'zombie' cars driving round cities in circles, prowling for passengers and never stopping in order to avoid incurring parking fees.

Traffic will still have to halt at intersections every now and then if pedestrians are ever going to be able to cross the road. Not every pedestrian crossing can have a bridge or an underpass.

Nevertheless, machine-driven cars will be more efficient consumers of road space than human drivers. Traffic conditions are not fixed fates which once imposed can never improve. A congestion charge has significantly reduced traffic flows in London, and the switch to almost-silent hybrid taxis has made walking the streets of Manhattan an even more inspiring experience than it used to be.[44] In any case, more efficient road use is not required to justify the introduction of self-driving cars. The horrendous death and injury toll imposed by human drivers is sufficient, together with the liberation from the boredom and the waste of time caused by commuting.

13.2.9 Other Affected Industries

Automotive cover represents 30% of the insurance industry, so a shift to self-driving cars will have a major impact on that industry. The most obvious effect should be a sharp reduction in pay-outs because there will be far fewer accidents. This in turn should mean far lower premiums: bad news for the insurance companies, good news for the rest of us.

Who will take out the insurance policy? When humans drive cars we blame them for any accidents, so they pay for the insurance. When machines drive, does the buck stop with the human owner of the vehicle, the vendor of the self-driving AI system, or the programmer who wrote its code? If the insured parties are Google and a handful of massive competitors, then the negotiating position of the insurance companies will deteriorate sharply from the present situation where they are 'negotiating' with you and me.

Warren Buffet ascribes some of his enormous success as the world's best-known investor to his decision to avoid areas he does not understand, including industries based on IT. He has massive holdings in the insurance industry. Unfortunately for him, software is 'eating the world',[45] and a large chunk of the insurance industry is about to be engulfed in rapid technological change. Buffet acknowledges that when self-driving cars are established, the insurance industry will look very different, almost certainly with fewer and smaller players.[46] It is very hard to say which of today's players will be the winners and losers.

The law of unintended consequences means that we cannot say how the insurance risks will change. Let's hope this never happens, but what if a bug – or a hacker – caused every vehicle in a particular city to turn left suddenly, all at the same time? How does an insurance company estimate the probability of such an event and price it? Important issues like this – and the ethical questions we will discuss subsequently – might slow down the introduction of self-driving cars. But they will not stop it. They are capable of resolution, just as we resolved questions about who would build the roads and who would have the right of way in different traffic situations in the decades after the first cars appeared.

People working in insurance companies will certainly not be the only ones affected by the move to self-driving cars. Machines will presumably be programmed not to violate local parking restrictions – and they will have no need to do so. That will remove a significant source of income from local authorities: parking charges generate well over $300 million a year for the city government of Los Angeles.[47]

Automotive repair shops will still be needed, but their business will shrink as it becomes restricted to maintenance and repairs necessitated by age rather than accidents. Happily, something similar can be said of doctors and nurses.

13.2.10 Programming Ethics

Imagine that your self-driving car is travelling down the road, minding its own business, when a child, unpredictably, dashes across the street ahead of you. Calculating at super-human speed, it analyses the only three available options: maintain direction, turn right or turn left. Even though it has already applied the brakes far quicker than any human could have, it forecasts (correctly, of course) that these options will result in the death, respectively, of the child, of an innocent adult bystander, or of you, its passenger.

Which option should it select? The question will have been answered in advance, even if only by default.

With grim humour, some have suggested that the answer will vary by car. Perhaps a Rolls Royce will always choose to preserve its owner, while a Lada may accord its occupants less respect.

What is happening here is the extension of human control over the world: the arrival of choice. Today, 27% of the victims of accidents are pedestrians and cyclists. What happens to them and the drivers of the cars which hit them is currently decided by the skill of the driver and blind chance. In future, we will have the power to affect it, and with increased power comes increased responsibility.

13.2.11 Driving Jobs

Clearly, self-driving vehicles will have a huge impact on society, sometimes in surprising ways. What impact will they have on employment? Are they indeed the poster child for technological unemployment?

There are 3.5 million truck drivers in the United States alone,[48] 650,000 bus drivers[49] and 230,000 taxi drivers.[50] How many of these jobs will be lost to machines?

It seems inevitable that machines will drive commercial vehicles. Articulated lorries are driven by professional drivers who are usually trained and who get lots of practice. Their backgrounds are checked and whose working hours and conditions are regulated. They cause fewer accidents per mile driven than cars owned by the rest of us. But because they are heavier, when they are involved in accidents, they cause much more damage to life and property. It is inconceivable that we will continue to allow humans to do this job for which machines are clearly better suited.

But driving is not the whole of the job. The people who drive trucks, delivery vans, buses and taxis have to deal with the myriad surprises which are thrown at them by life, which is an untidy business at best. If a consignment of barbed wire falls off the back of a truck in front, they will get out and help. They are also often responsible for loading and unloading their vehicles.

Sceptics about technological unemployment point out that planes have been flying by wire for decades, with human pilots in control for only around three minutes of an average commercial flight. We have yet to dispense with the services of human pilots.

However, a 747 is very different to a truck travelling down Highway 66. A truck is an expensive vehicle, and capable of inflicting severe damage, but commercial planes are on a different scale: they cost many millions of dollars each, and their potential to cause harm was graphically and tragically demonstrated in New York in 2001. Furthermore, those three minutes of human control are in part due to the difficulty of resolving the edge cases we discussed before. In road vehicles, if not in planes, we are well on the way towards resolving those.

Consider the process of delivering a consignment from a warehouse to a supermarket or other large retail outlet. Amazon's fleet of Kiva robots show that warehouses are well on the way towards automation. The unloading bays at the retail end are also standardised for efficiency: a system which automates the entire unloading process from a truck into the retailer's receiving area is technically feasible today, and with the exponential improvement in robotics and AI, it won't be long before it is economically feasible as well.

As we have seen, robots are becoming increasingly flexible, nimble and adaptable. They can also increasingly be remotely operated. Most of the situations a driver could deal with on the open road will soon be within the capabilities of a robot which does not need sleep, food or salary. On the rare occasion when human intervention is needed, the gig economy can probably furnish one quickly enough.

Once it is economically feasible to replace human drivers with machines, it is a very short step to it being economically compelling. As mentioned previously, drivers account for 25%–35% of the cost of a trucking operation.[51] You can't escape the invisible hand of economics for long. In a free market, once one firm replaces its drivers, the rest will have to follow suit or go out of business. Of course, trade unions and sympathetic governments may try to stop the process in some jurisdictions. They may succeed for a while, but only by rendering their industry uneconomic, and burdening their customers with unnecessary costs which will damage them in turn.

Other governments will take a different approach, and the competitive disadvantage imposed by resistance to change will become apparent. It will not be manifest only in the case of truck drivers, but in all areas of the economy, and regions and countries which do resist will find their living standards declining fast. Over time, it will prove unsustainable.

In July 2017, India's transport minister said 'we won't allow driverless cars in India. I am very clear on this. In a country where you have unemployment, you can't have a technology that ends up taking people's jobs'. Opposition politicians rolled their eyes and predicted the policy would be swiftly reversed, and even the minister himself gave himself an escape route: 'Maybe some years down the line we won't be able to ignore it, but as of now, we shouldn't allow it'.[52]

In fact, the process has already started. In the Yandicoogina and Nammuldi mines in Pilbara, Western Australia, transport operations are now entirely automated, supervised from a centre in Perth, which is 1200 miles away.[53] Mining giant Rio Tinto was prompted to take this initiative by economics: the decade-long mining boom caused by China's enormous appetite for raw materials. Drivers earned large salaries in the hazardous and inhospitable environments of these remote mines, which made the investment case for full automation irresistible.[54] The economics are going in the same direction everywhere – fast.

13.2.12 Road Rage against the Machine

The automation of driving will have a major impact on the overall job market. Truck and delivery driver is the most commonly named occupation in 29 US states (57% of them).[55] These drivers and their unions are unlikely to surrender their jobs without a struggle, and there may even be violence – call it road rage against the machine. Perhaps it will only be violence against property, with the cameras and sensors of self-driving vehicles being spray-painted over. Perhaps there will be violence against the owners of fleets which adopt self-driving vehicles.

Much depends on how governments handle the transition. At the moment, very few of them are even considering the issue. In places like Pittsburgh, which has become a centre for self-driving technology, they welcome the new jobs that have been created. They are great jobs: highly paid employment for PhDs, plus a few jobs for the chaperones who sit in the test cars and take over driving duties every now and then. The people of Pittsburgh can see that these will not replace all the driving jobs that will be lost, but the politicians have no answers.[56]

Of course, it won't only be truck drivers. In 2014, the medallion that gives you the right to drive a yellow taxi in New York City sold for $1.3 million. Drivers and investors took out huge loans to buy them, but their price has crashed as Uber, Lyft and others have entered the market. Many people who thought they had achieved financial independence now find themselves bankrupt, or working ever harder to meet interest payments on the loans.

This is just a foretaste of what is to come. When self-driving cars become a common sight on roads around the world, they will be like the canary in the coal mine which gave miners a bit of advance warning that they were in great danger. Self-driving vehicles will have the effect of alerting everyone else to the prospect of widespread technological unemployment.

13.3 WHO'S NEXT?

Carl Benedikt Frey and Michael Osborne are the directors of the Oxford Martin Programme on Technology and Employment.[57] In 2013, they produced a report entitled *The Future of Employment: How Susceptible Are Jobs to Computerisation?* It has been widely quoted, and its approach to analysing US job data has since been used by others to analyse job data from Europe and Japan. You can read more about this report in the appendix.

The report analysed 2010 US Department of Labour data for 702 jobs, and in a curious blend of precision and vagueness, concluded that '47% of total US employment is in the high-risk category, meaning that associated occupations are potentially automatable over some unspecified number of years, perhaps a decade or two'. Nineteen per cent of the jobs were found to be at medium risk and 33% at low risk. Studies which have extended these findings to other territories have yielded broadly similar results.

The report suggested that the automation would come in two waves. 'In the first wave, we find that most workers in transportation and logistics occupations, together with the bulk of office and administrative support workers, and labour in production occupations, are likely to be substituted by [machines]'.[58] It makes intuitive sense that lower income jobs would be less cognitively demanding, and hence easier to automate.

13.3.1 Lower Income Service Jobs

Arriving at JFK airport is a very mixed blessing. It is always exciting to visit New York City, but immigration control at JFK often seems to be run by sadists offering a main dish of rudeness and a side order of incompetence. How nice would it be to disembark from a plane after a long-haul flight and wait in comfort (maybe even seated!) while robotic baggage handlers fetch your luggage, and unintrusive scanning drones quickly and efficiently check your face against the documents you have transmitted. While JFK is reported to be getting a $10 billion overhaul, airports further east are already competing to provide much better experiences, since that is where the growth markets are.[59] In the medium term, we can expect to be processed by many fewer humans at airports, and I suspect we will not regret it.

As you read this, there are people all over the world inspecting pipes and cables. This is lonely, often boring, and occasionally dangerous work, but it has to be done. Not by humans for very much longer, though. Smart drones are already taking over in some industries, and much of the rest will follow.[60] In a decade or two, the Wichita lineman won't be still on the line.

13.3.1.1 Retail

Automation is not new to the retail industry. ATMs took over the job of dispensing cash in banks years ago, and self-service checkouts are familiar sights in supermarkets. Neither is it new in food service specifically. Few people in major cities now buy their sandwiches bespoke from a human who prepares their food in front of them. Far more buy their lunch in ready-made packages.

Back in 1941, the Automat chain served half a million American customers a day, dispensing macaroni cheese, baked beans and creamed spinach through cubby-holes with glass doors.[61] The chain declined in the 1970s with the rise of fast-food restaurants serving better-tasting food, such as Burger King and of course McDonalds. But these chains themselves are now discovering the economic appeal of automation.

Chili's Bar and Grill is rolling out a tablet ordering system, and Applebee's began delivering tablets to all its 1,800 restaurants in 2014.[62] There has been heated political debate about whether these and similar initiatives are prompted by increases in minimum wage levels, but the truth is that AI systems and robots are rapidly becoming more efficient at cost levels which humans can never hope to match.

13.3.1.2 Customer Preference

Cost saving is not the only reason for this kind of automation. In many situations, humans prefer to transact with machines rather than other humans. It can be less time-consuming, and require less effort. It can also make a service available for longer hours, perhaps 24 hours a day. Bank ATMs are the classic example. Another are the automated passport control systems now installed at many airports, which many people opt to use in preference to the manned channels.

A report published in April 2015 by Forrester, a technology and market research company, claimed that 75% of procurement professionals and other people buying on behalf of businesses (i.e. B2B buyers) prefer to use e-commerce and buy online rather than deal with a human sales representative. Once the buyers have decided what they want, the percentage rises to 93%.[63] Forrester pointed out that many vendors were ignoring this fact and obliging customers to speak to a human. This is no doubt partly at least because human sales people are currently much better able to up-sell the buyer, but this is also one of the reasons why buyers prefer e-commerce. Forrester argued that companies which wait too long to offer good e-commerce channels risk losing market share to more digitally minded competitors.

13.3.1.3 Call Centres

We are still at the very early stages of introducing AI to call centres. For many of us, dealing with call centres is one of the least agreeable aspects of modern life. It normally involves a good deal of waiting around, listening to uninspiring hold music, followed by some profoundly unintelligent automated routing, and finally a conversation with a bored person the other side of the world who is reading from a script written by a sadist.

One of the leaders in introducing genuine AI to call centres is Swedbank, one of Sweden's biggest banks, with 9.5 million customers and 160,000 employees. It has 700 people working in contact centres, which handle two million customer calls each year. It has worked with the American software company Nuance to introduce a basic AI called Nina,[64] which learns what customers want and how best to help them by assimilating searches made on the company website and enquiries made at the contact centres.[65] In December 2015, Nina was handling 30,000 calls a month, and taking care of many of the straightforward transactional calls – like transferring money from one account to another – which were previously clogging up the call centres. The aim is to free up the agents in the contact centres to concentrate on more complicated activities, like taking out a mortgage. But even taking out a mortgage isn't rocket science. Given exponential progress, if Nina can handle transfers today, it will surely be able to handle mortgage applications before long.

13.3.2 Manual Work

Occupations requiring physical labour will take longer to automate than clerical and administrative jobs because getting robots to be dextrous and flexible is surprisingly hard. As we saw in Section I, progress is rapid, but much remains to be done. Manual work on routine, repetitive tasks on assembly lines will continue to see machines taking over, but physical labour in unstructured environments like building sites will remain the preserve of human workers for a while longer.

Agriculture was the poster child for automation (or rather, mechanisation) during the Industrial Revolution. The process continues into the age of more cognitive automation.

Harper Adams University is a specialist provider of higher education for the agricultural sector in the United Kingdom, and in 2017 they managed to plant, tend and harvest an acre and a half of barley using only autonomous vehicles and drones.[66] The technology to do this is not all generally available yet, and it is not cheap. But it will be.

A company called Octinion claims its robot is cost-competitive with humans at picking strawberries grown on tabletop systems rather than in fields, which it claims to be the direction the industry is heading. The robot picks one berry every five seconds whereas a human manages one every three seconds, but the robot costs much less to run after the initial purchase.[67]

Manufacturing accounts for over a third of China's GDP, and employs more than 100 million of its citizens. Historically, China's competitive strength in manufacturing has been its low wage costs, but this is changing fast: wages have grown at 12% a year on average since 2001, and Chinese manufacturers are embracing automation enthusiastically. China is now the world's largest market for industrial robots, but it has a long way to go before it catches up with the installed base in more developed countries.

Industrial robots are far from perfect, and manufacturers have underestimated the progress still required. In 2011, the CEO of Foxconn, a $130 billion -turnover Taiwanese manufacturer that is famous for making iPhones, declared a target of installing a million robots by 2014. The robots failed to perform as he hoped, and the actual installation rate has been much slower. But the robots are improving fast.[68]

YouTube now has plenty of videos showing how far automation has advanced in warehouses, but people still point to Amazon's ravenous hiring of warehouse people as proof that 'It's a myth that automation destroys net job growth', to quote Dave Clark, a senior executive at operations there.[69] Unfortunately it is not a myth, as we have seen. Amazon has 100,000 robots in its warehouses, but it is hiring humans because it is growing fast and the robots cannot yet do everything the humans can. Amazon's no-profits business model and its brilliant management means that it is eating the lunches of bricks and mortar retailers, and soon it will eat their dinners too. Global online retail sales are only expected to be 8.8% of total retail sales in 2018, but they are growing fast – they are up from 7.4% in 2016.[70] (In the United Kingdom, which decided long ago that it prefers congestion to new roads, they are 15.6% and still growing fast.)

Amazon is the primary beneficiary of this massive structural shift. To rely on it to keep humans in work is foolhardy. One of the main jobs still held by humans in its warehouses is picking items out of boxes, a task which is hard for robots thanks to Moravec's paradox. Every year Amazon holds a competition for manufacturers of robotic pickers; the first company which creates a human-level picker will receive $250,000, Amazon will stop using so many humans, and people will finally have to stop claiming that Amazon's warehouses prove that automation does not create widespread lasting unemployment. Alberto Rodriguez, a roboticist at MIT who has worked with Amazon on the problem, thinks it will take five years or so.[71]

13.3.3 The Professions

It is certainly not only low-paid, relatively low-prestige service jobs that will be automated. The professions are vulnerable too: lawyers, doctors, architects and journalists. Sometimes accused of being conspiracies against lay people, these are protected occupations, with demanding entry requirements and restrictions on the number of trainees who can join the professions each year. They have commanded prestige and high salaries, but that may be about to change.

13.3.3.1 Journalists

Nuance, the company behind Swedbank's Nina call centre AI, offers services for journalists, helping them create interviews and articles faster. But Narrative Science, a company established in Chicago in 2010, has an AI system which writes articles without human help. Called Quill, it already produces thousands of articles every day on finance and sports for outlets like Forbes and Associated Press (AP).[72] Most readers cannot identify which articles are written by Quill and which by human journalists, and Quill is much faster.

Quill starts with data – graphs, tables and spreadsheets. It analyses these to extract particular facts which could form the basis of a narrative. It then generates a particular plan, or narrative for the article, and finally it crafts sentences using natural language generation software.

A British company called Arria offers the same functionality, but sells mainly to corporations trying to make sense of the tsunami of data which threatens to overwhelm them.[73]

Quill has not rendered thousands of journalists redundant. Instead, it has sharply increased the number of niche articles being written. Newspaper revenues have declined sharply since the turn of the century, as classified ads for jobs, houses and cars migrated online. News services like AP increased the daily quota of articles for each journalist, cut back the number they employed, and reduced the number of articles they produced on the quarterly earnings reports of particular companies, for instance. Quill and similar services have enabled them to reverse that decline. AP now produces articles on the quarterly reports of medium-sized companies that it gave up covering in such detail years ago.

Kristian Hammond, founder of Narrative Science, forecast in 2014 that in a decade, 90% of all newspaper articles would be written by AIs. However, he argued that the number of journalists would remain stable, while the volume of articles increased sharply. Eventually, articles could become tailored for particular audiences, and ultimately for each of us individually. For instance, an announcement by a research organisation that inflating your car tyre correctly could reduce your spend on petrol by 7% could be tailored – perhaps with the help of your Digital Personal Assistant – to take into account your particular car, the number of miles you drive each week, and even your style of driving. (Although of course by then you will perhaps not do the driving yourself anyway.)

The prediction that the number of human journalists will remain stable sounds reassuring, and indeed you would expect someone marketing an automating technology to say that. Given the exponential improvement of AIs, it is a brave prediction.

TV presenters should also be feeling nervous. In December 2015, Shanghai's Dragon TV featured Xiaoice (pronounced Shao-ice), an AI weather presenter with a remarkably life-like voice,[74] based on the Mandarin version of its Cortana digital assistant software. Audience feedback was positive.[75]

13.3.3.2 Other Writers

Not everyone who spends their working days crafting crisp sentences is a journalist. They might be PR professionals or online marketers, for instance. A company called Persado claims that marketing emails drafted by its AI have a 75% better response rate than emails written by human copywriters.[76] Citibank and American Express are customers as well as investors.

(Some people think that AI will make most marketers redundant as it takes more and more purchasing decisions on our behalf, and brands thus become far less valuable and important. I'm not so sure, but it's an interesting thought.[77])

In January 2016, a researcher at the University of Massachusetts announced an AI which can write convincing political speeches for either of the two main US political parties. The system learned its craft by ingesting and analysing 50,000 sentences from congressional debates.[78]

Two other professions have come under particular scrutiny with respect to their susceptibility to machine automation: the law and healthcare. Let's look at these in turn.

13.3.4 Lawyers

Whatever Hollywood thinks, most lawyers do not spend their days pitting their razor-sharp wits against equally talented adversaries in front of magisterial judges, eliciting gasps of admiration from around the courtroom as they produce the winning argument with a flourish. Most of the time they are reading through piles of very dry material, looking for the thread of evidence which will convict a fraudster, or for the poorly drafted phrase which could undermine the purpose of a contract.

13.3.4.1 Discovery

Many lawyers get a lot of their on-the-job training through the 'discovery' process. Known as 'disclosure' in the United Kingdom, this is a pre-trial process in civil law in which both sides must make available all documents which may affect the outcome of the case. An analogous process takes place in the 'due diligence' phase of a corporate merger and acquisition (M&A), in which teams of junior lawyers (and accountants) spend weeks locked away in data rooms, reading through material which can run into millions of documents, looking for something which would clinch the case, or, in the case of M&A work, provide a reason to terminate or renegotiate the deal.

Looking for a needle of fact in a haystack of paper is work more suited to a machine than a man. And although lawyering is a very conservative profession, there are signs that it understands what is coming better than some others. RAVN Systems is the British AI company behind an AI system called Ace, which reads and analyses large sets of unstructured, unsorted data. It produces summaries of the data, and highlights the documents and passages of most interest according to the pre-set criteria.[79] When one of the United Kingdom's largest law firms started working with Ace, it was regarded as pioneering, experimental and somewhat risky. Two years later that law firm was promoting its own services to potential new clients on the basis that it knew best how to exploit the advantages of RAVN Ace. Bear in mind that two years is a very short time for anything at all to happen in the legal industry!

Typically, a new client's data will present a new set of challenges. It usually takes a few days to train the system how to read the data, which currently involves human intervention. Once the training is complete, the work proceeds without human involvement, and the system will finish the work much faster than human lawyers could. This means that law firms are having to work out new ways of billing their clients: the old system of hourly rates is under challenge.

13.3.4.2 Revealing the Iceberg

Forward-thinking lawyers are actually excited about the arrival of this sort of automation. Rather than fearing that it will destroy the jobs of junior lawyers, making it impossible for young people to learn the profession, they believe it will increase the amount of cases that can be handled. To illustrate this, imagine a large supermarket chain that wants to know the implication of making a small change to the employment contracts of all its in-store employees – tens of thousands of people. Previously, its employment law firm would have said that this task could not be undertaken cost-effectively with any degree of rigour. RAVN Ace and systems like it make this kind of work possible, opening up whole new avenues of work for law firms. It is like standing nervously on a body of ice, thinking you are only separated from freezing water by a thin layer, and suddenly discovering that in fact you are standing on an iceberg, with a huge mass of previously unknown solid ice beneath your feet.

As Greg Wildisen, MD of Neota Logic, a firm providing an AI platform for lawyers, puts it, 'So many legal questions go "un-lawyered" today that there is enormous scope to better align legal resources through technology rather than fear losing jobs'.[80]

So, in the short and medium term, machine automation of white-collar jobs opens up vast new areas of work that can be undertaken, and doesn't throw the incumbent humans out of work. They are still needed to train the system at the start of a large new assignment, and to process more complicated documents.

But as RAVN Ace and its successors improve – at an exponential rate, of course – they will be able to take on more and more of the sophisticated and demanding aspects of the lawyers' work. No one can be absolutely sure yet whether this process will hit a wall at some point, leaving plenty of work for humans, or whether it will continue to the point where there are very few jobs left for humans. My own view is that within a few short decades, the machines are coming for most of our jobs.

The short-term explosion of work which happens as the iceberg of latent demand is revealed can give us a false sense of security. The phenomenon of automation leading to job creation is sometimes called the automation paradox.[81] But the paradox may turn out to be short-lived.

13.3.4.3 Forms
Another fairly basic form of legal work is the completion of boilerplate (standard) forms to establish companies, initiate a divorce, register a trademark, request a patent and so on. A company called LegalZoom was established in 2001 to provide these services online, and increasingly, to automate them. LegalZoom now claims to be the best-known law brand in the United States,[82] and in 2014, the private equity firm Permira paid $200 million to become its largest shareholder. Another company, Fair Document, helps clients complete forms for less than $1,000, one-fifth the amount it would have previously cost.[83]

13.3.4.4 More Sophisticated Lawyering
At the other end of the spectrum from the 'grunt' work of discovery and filling out legal forms, one of the most sophisticated and important jobs that senior and successful lawyers are asked to undertake is to estimate

the likelihood of a case winning. The advice is vital as it will determine whether large amounts of money are spent. A team led by Daniel Martin Katz, a law professor at Michigan State University, developed an AI system that analysed 7,700 US Supreme Court cases. It predicted the verdicts correctly 71% of the time.[84]

Another job for experienced lawyers in common law jurisdictions such as the United States and the United Kingdom is identifying which precedent cases to deploy in support of litigation. A system called Judicata uses machine learning to find the relevant cases using purely statistical methods, with no human intervention.[85]

In September 2016, the eastern Chinese province of Jiangsu started using 'legal robots' to review cases and advise on sentencing. In the following year, they reviewed 15,000 cases, many of them traffic violations. They detected errors in more than half the cases.[86]

Will entire sections of the legal industry be automated in the next few years or decades? How about patent lawyers, for instance? Senior patent lawyers are highly skilled and articulate people, but much of the work involved in securing a patent is routine and could perhaps be automated. In November 2015, I took part in a debate at the IMAX cinema in London's Science Museum. The motion was 'This House believes that within 25 years, a patent will be applied for and granted without human intervention'. Patent lawyers comprised a good part of the audience, and although the motion was vigorously opposed by two senior patent lawyers, the motion was passed. Not exactly turkeys voting for Christmas, but certainly food for thought.

13.3.5 Doctors

Doctors are a scarce resource. Only bright and dedicated people are admitted to the relevant university and post-graduate courses, and these courses demand many years of hard study. Hospitals and local surgeries are organised to maximise the availability of this resource, but some critics argue that they are organised for the benefit of the doctors rather than the patients. In 2015, senior doctor and medical researcher Eric Topol published a book called *The Patient Will See You Now*, which he argues should become the mantra for the profession, replacing the current one, which he says is 'the doctor will see you now'.

Suggesting acerbically that the initials MD stand for medical deity, Topol accuses many doctors of being arrogant and paternalistic towards their patients, assuming they are unable to understand the detailed information regarding diagnoses, and withholding information from patients so as not to upset them. He believes that the digital revolution will start to overturn this unsatisfactory state of affairs, as it will place cheap and effective diagnostic tools in the hands of patients.

13.3.5.1 Better and Cheaper Diagnostics

In April 2016, researchers at Indiana University announced that a test of open-source machine learning algorithms on 7,000 free-text pathology reports from 30 hospitals yielded equal or better diagnoses than humans had made. The computers were also faster and cheaper.[87]

This sort of technology will become much more widely available. A British start-up called Babylon charges customers £5 a month for phone (and videophone) access to a dedicated team of doctors. Before a doctor comes on the line, the patient is triaged by a machine.[88] As AI improves, the role of the human doctor in this process will continuously be reduced.

Smartphones are increasingly able to gather medical data about us, and perform basic analysis. By attaching cheap adapters to their phones, patients can quickly take their blood pressure, sample their blood glucose and even perform an electrocardiogram. Your breath can be sampled and digitised, and used to detect cancer or potential heart problems. Your camera's phone can help screen for skin cancers. Its microphone can record your voice, and that data can help gauge your mood, or diagnose Parkinson's disease or schizophrenia.

All this data can be analysed to a certain level within the phone itself, and in many cases that will suffice to provide an effective diagnosis. If symptoms persist, or if the diagnosis is unclear or unconvincing, the data can be uploaded into the cloud-that is, to server farms run by companies like Amazon and Microsoft. The heart of diagnosis is pattern recognition. When sophisticated algorithms compare and contrast a set of symptoms with data from millions or even billions of other patients, the quality of diagnosis can surpass what any single human doctor could offer.

Ross Crawford and Jonathan Roberts are professors of orthopaedic research and robotics respectively at Queensland University of Technology. In an article in January 2016,[89] they argued that doctors need to understand that diagnostic services can be made available more cheaply with the assistance of machine intelligence, and reach all the patients who need them, not just those in rich countries who are already manifesting symptoms.

They don't think this will render doctors unemployed. As with the law, there is an iceberg of unmet healthcare needs – needs which automation and machine intelligence can satisfy. Formed in Mumbai in 1996, Thyrocare Technologies is the world's largest thyroid testing laboratory. Its founder, Dr A Velumani, had the insight that 90% of people who could benefit from diagnostic tests were not receiving them because they were too expensive, so the tests were restricted to those already manifesting symptoms of disease. He established Thyrocare to address this latent demand, and now it processes 40,000 samples a day.[90]

As this iceberg is revealed, the healthcare industry – like the legal industry – can perform a far better job by reaching many more people at greatly reduced average cost. At first, there will be just as much need for doctors as before: they will continue to carry out the more sophisticated diagnoses, while machines (possibly deployed by less highly trained people) deliver the routine work. But as we keep observing, the machines are getting smarter at an exponential rate. In time, what is to stop them performing the doctors' other roles as well, and doing them better, faster and cheaper?

13.3.5.2 Prescribing

If machines can diagnose, can they go on to prescribe, and to fulfil prescriptions? The University of California in San Francisco has installed a robot pharmacist which is reported to have prepared 6 million prescriptions with only one error – a track record which is 60,000 times better than human pharmacists.[91]

13.3.5.3 Keeping Current

As we saw in Section 3.4, machines are reaching parity with humans in pattern recognition, and will quickly become much better. They are already much better than human professionals at keeping up with new developments in their field.

A human doctor would have to read for 160 hours a week just to keep up with the published medical research. This is clearly impossible for a human, but machines have no such bandwidth restriction. IBM is pushing aggressively into the medical industry with its Watson AI system. According to Samuel Nessbaum of Wellpoint, a private healthcare company, Watson's diagnostic accuracy rate for lung cancer is 90%, which compares favourably with 50% for human physicians.[92]

IBM has been criticised for pretending that Watson is a unitary system rather than a kludge of different systems which can be mixed and matched according to need. It is also accused of scaling back its ambition by tackling much smaller projects than the 'moonshots' it was originally earmarked for, like curing cancer.[93] Kris Hammond, the founder of Narrative Science whom we met when discussing journalists, says that 'everybody thought [winning *Jeopardy*] was ridiculously impossible, [but now] it feels like they're putting a lot of things under the Watson brand name – but it isn't Watson'.[94] In March 2016, DeepMind founder Demis Hassabis went as far as to say that Watson is essentially an expert system as opposed to deep learning one.[95]

IBM is unfazed by this kind of criticism. It says that Watson is now being used by hundreds of companies to solve particular problems – companies like the Australian energy group Woodside, which used it to review 20,000 documents from 30 years of engineering projects to identify, for instance, the maximum pressure that a certain type of pipeline can withstand. It might be a form of marketing sleight of hand to apply the Watson brand to all these applications, but the company spent a great deal of time and money to create that brand, and it would be unreasonable to expect it not to try and recoup that investment.

That said, IBM is developing a new brand for its commercial AI offering. Celia stands for Cognitive Environments Laboratory Intelligent Assistant, and it seems to be a more user-friendly front end, enabling business analysts, for instance, to interact with it by speech, and by manipulating virtual objects in an augmented reality field.[96]

And IBM is still pursuing moonshots, in the medical field and elsewhere. As we have noted several times, machine learning is fuelled by data. In October 2015, IBM paid $1 billion for Merge Healthcare, a company with 30 billion medical images,[97] and $2 billion for the digital assets of The Weather Company, to build a weather forecasting service. At the end of the year, it unveiled Avicenna, a product of the Watson healthcare business unit designed to help radiologists prioritise which images to review, and help them make diagnoses.[98] The interesting question is, how long before at least some of those radiologists turn out to be superfluous to the process. Two of the world's leading AI researchers, Geoff Hinton and Andrew Ng, have both said that radiologists training today will have truncated careers.[99]

13.3.5.4 Operations

You might think that the hands-on physical and frankly messy business of surgical operation will be undertaken by humans rather than machines for the foreseeable future. Probably not. One of the most highly skilled professionals in the emergency suite is the anaesthetist, and Johnson & Johnson has an automated version called Sedasys which, despite fierce opposition from the profession, has FDA approval to provide the anaesthesia in less challenging procedures like colonoscopies. It has carried out thousands of operations in Canada and the United States.[100] In March 2016, Johnson & Johnson announced that it was exiting the Sedasys business due to sluggish sales, despite the machine costing $150 per operation whereas a human anaesthetist costs $2,000.[101] This will certainly not be the last setback in the progress of the machines, but in the long run the economic facts will prevail – although perhaps more slowly in industries where the normal rules of the market do not always apply.

In May 2016, an academic paper announced that a robotic surgeon had out-performed human peers. The Smart Tissue Autonomous Robot (STAR) operated on pig tissue and did the job better, although four times more slowly, than humans operating alone – and also better than humans aided by the semi-robotic Da Vinci system.[102]

In September 2017, Hong Kong's newspaper of record reported that the world's first successful automated dental implant surgery had fitted two new teeth into a woman's mouth. The patient looks terrified in the article's picture, but it was her first time as well as the world's.[103]

A study published in the *Lancet* medical journal of the 20,000 British men who had prostate surgery in the first half of this decade reported that they were choosing to have their operations in clinics which used robots. This preference was strong enough to force some of the clinics without robots to close.[104]

13.3.5.5 Education

Teachers are the active ingredient in education – at school level, anyway. Studies have shown repeatedly that the quality of teaching makes an enormous difference to how well a student performs at school and afterwards. But schools cannot afford enough of them, governments burden them with bureaucracy, and most countries' cultures undervalue them.

What happens when the learning of every pupil is monitored minutely by AI? When every question she asks and every sentence she writes is tracked and analysed, and appropriate feedback is provided instantly? Teachers will play the role of coach instead of instructor, but as with the other professions, their scope for contribution will shrink.

The beachhead for AI in education is marking, also known as grading. This is the bane of many teachers' lives, and they will welcome an assistant which can relieve them of the duty. A company called Gradescope marks the work of 55,000 students in 100 US universities, marking simple, multiple-choice types of test. It raised $2.6 million in April 2016 to develop its product into complex questions and essays.[105] Large corporates like Pearson and Elsevier which provide education services are moving in the same direction.

Towards the end of 2015, 300 students at Georgia Institute of Technology were, unbeknownst to them, guinea pigs in an experiment to see whether they would notice that one of their nine teaching assistants was a robot. Only ever in contact via email, they would ask questions like 'Can I revise my submission to the last assignment?' and receive answers back like 'Unfortunately there is not a way to edit submitted feedback'. None of the students noticed that Jill Watson, named after the IBM Watson system 'she' ran on, was in fact an AI.[106]

13.3.6 Financial Services

The finance sector is an obvious target for machine intelligence, with high-value (and high-priced!) services based on vast amounts of data. Human equity analysts and brokers will increasingly struggle to provide value in the face of competition with machines which can ingest all the relevant data, and never forget any of it. The provision of advice to investors is also migrating to machines, with systems like SigFig incorporating a client's risk appetite and investment style into its algorithms' analysis of low-cost opportunities and recommendations.[107] Similar so-called 'robo-adviser' services are available from Betterment, Wealthfront and Vanguard.[108]

What is known as 'fintech' is one of the hottest areas for venture capital (VC) investment at the time of writing, and banks are spending considerable amounts of time and energy on working out where the most powerful disruption to their business models will come from, and whether they can do the disrupting themselves rather than be its victims. Banking, especially retail banking, has traditionally been a conservative, slow-moving industry, but the pace is picking up. Goldman Sachs reports that 40% of all US cheques are now processed electronically, despite that service being only 4 years old.[109]

In September 2017, the CEO of Deutsche Bank made headlines when he told an audience that 'in our bank we have people doing work like robots. Tomorrow we will have robots behaving like people. It is going to happen. The sad truth for the banking industry is, we won't need as many people as today'.[110]

The CEO of Mashreq Bank in Dubai agrees. 'By using artificial intelligence, employment at banks will shrink over time', he said, adding that his own bank would shed 10% of its 4,000 people in the next 12 months. The average cost of its junior employees is Dh250,000, and the cost of replacing them with AI is a one-time investment of Dh30,000, he added. The kinds of services that can now be done with AI include issuing new credit cards, depositing checks and opening new accounts.[111]

The CEO of UBS said that his bank could reduce headcount from nearly 100,000 to around 70,000 over the next few years. It does not sound very comforting to hear that the bank 'can have 30% [fewer people], but the jobs are going to be much more interesting jobs, where the human content is crucial to the delivery of the service'.[112]

13.3.6.1 Trading and Execution
These services deploy primitive forms of AI at the moment. According to market research firm Preqin, thousands of hedge funds, managing $200 billion of assets, use computer models in most of their trades. But they are using traditional statistical methods rather than AI which learns and evolves. This is changing. Bridgewater, the world's largest hedge fund, hired David Ferrucci away from IBM, where he had project managed the development of the version of Watson which beat Ken Jennings at *Jeopardy*.[113] January 2016 saw the inaugural trades of Aidyia, a hedge fund based in Hong Kong whose chief scientist is Ben Goertzel, one of the leading researchers in artificial general intelligence (AGI).[114]

The AHL hedge fund started using AI seriously in trading in 2014, and within a year it was generating half of the division's profits. With $5 billion under management, AHL is one of the largest parts of the Man Group, which in turn is the world's largest publicly traded hedge fund, managing $96 billion.[115] Nick Granger, the executive who persuaded the company to trust the AI, became AHL's CEO. 'The idea that the humans will just disappear … is just not right. It's just that they move to different tasks, to higher value-added tasks. We need smarter humans than we did'. This begs the question of what happens to the existing staff when the bank has hired lots of these smarter humans. Luke Ellis, the CEO of the parent Man Group is less sanguine: 'My hope is always that there will be parts that humans do that AI doesn't do, [but] I wouldn't bet my life on it'.[116]

Other experts agree. 'The human mind has not become any better than it was 100 years ago, and it's very hard for someone using traditional methods to juggle all the information of the global economy in their head', says David Siegel of Two Sigma, another hedge fund which uses AI. 'The time will come that no human investment manager will be able to beat the computer'.[117]

Jeff Tarrant of investment firm Protégé Partners thinks that the technology is in an early phase of adoption, and 'there is going to be mass unemployment in asset management in the next several years'.[118]

'Algo trading' has many critics in financial circles, who point out that they chase spurious correlations (such as the fact that divorce proceedings in Maine have consistently tracked sales of margarine), and that they can move markets in ways that are impossible to follow and are potentially dangerous. But in a financial world that has become so complex that mere humans can no longer follow it, they may not only be inevitable, but also necessary. David Siegel says 'People talk about how robots will destroy the world, but I think robots will save it'.

In July 2017, JP Morgan, the world's biggest investment bank, announced that a trial in Europe of a deep reinforcement learning system for the execution of trades had been successful, and the system, known as LOXM, would be rolled out to the United States and Asia. The execution of trades involves no selection of which equities to buy, but decisions about when to carry out the trades and in what volumes can significantly affect the price achieved, and were previously considered a skilled task for humans. JP Morgan claimed the move would give it a two-year advantage over rivals, which would cost them many millions to erase.[119]

Top analysts are among the highest-paid people in investment banks. They are targeted by a number of fintech companies like Kensho, which sorts through thousands of data sets to produce reports in minutes which would take skilled humans days. For instance, asking the system about the Syrian civil war will generate a report showing its impact on companies, currencies and commodities in as many countries as you like. Kensho's founder, Daniel Nadler, thinks between a third and a half of finance employees will be redundant within a decade.[120]

13.3.6.2 Compliance

One of the drivers of AI use by financial services firms is the ever-growing and increasingly complex web of compliance requirements imposed by governments and regulators. Systems like IPSoft's Amelia help insurance firms and other financial services companies to navigate this web and make sure the forms and procedures used by staff are up-to-date.[121]

Global banks are regularly fined hundreds of millions of dollars for carrying out illegal or sanctioned trades. Standard Chartered's regulatory costs rose 44% in 2015 to $447 million as it was obliged to hire thousands of additional staff to deal with compliance requirements. In March 2016, it announced a major investment in AI systems to oversee its traders' behaviour, and to match their activities against regulatory norms.[122]

It seems that managers in financial services are becoming aware of the potential threat to their livelihood. A survey of 1,700 managers in 17 different industries carried out in the autumn of 2015 by the consulting firm Accenture revealed substantial anxiety about automation. Overall, a third of the managers feared that intelligent machines threatened their jobs, with the level rising to 39% among senior managers. Unsurprisingly, the anxiety was highest (50%) among managers in the technology sector, but it was also high in banking, at 49%.[123]

13.3.7 Software Development

You might think that one of the safest jobs would be software development, but Google's AutoML project is one of a number that are trying to overcome the blockage caused by a shortage of machine learning developers by getting machine learning software to build machine learning software. CEO Sundar Pichai explained that '[cutting-edge AI systems] are hand crafted by machine learning scientists and literally only a few thousands of scientists around the world can do this … We want to enable hundreds of thousands of developers to be able to do it'.[124]

13.4 SCEPTICAL ARGUMENTS

13.4.1 The Reverse Luddite Fallacy

As we saw in Section 12.4, the Luddite fallacy is the argument that we have had automation for centuries and it has never caused widespread lasting unemployment, and therefore it never can. People have raised fears about unemployment many times down the years, and they have always proved false. They are like the boy who cried wolf in Aesop's fable.[125]

When the argument is expressed this baldly, its weakness is clear: it is akin to saying that we have never sent a person to Mars so we never will. Or to someone in 1902 saying that despite numerous false alarms, we have not achieved powered flight in the past, and therefore we never will.

It is also historically inaccurate – if you count horses as employees. As we saw previously, there were 21.5 million horses in America in 1915, mostly working on farms; now there are pretty much none. Even if you only count humans as employees, your definition of 'lasting' unemployment has to be extreme: the Engels Pause is the name for the fact that labour's share of national income was depressed, in the United Kingdom at least, for most of the first half of the nineteenth century.

As for crying wolf, it is worth remembering that at the end of the story, the wolf did turn up and kill all the sheep. Automation has been going on for centuries, and past claims that it was causing permanent widespread unemployment have been proved wrong. But we should not be complacent when there are good reasons to think that this time, it may be different.

Past rounds of automation replaced human and animal muscle power. That was fine for the humans who were displaced, as they could go on to do jobs which were often more interesting and less dangerous, using their cognitive faculties instead of their muscle power.

Intelligent machines are replicating and in many cases improving on our cognitive abilities. As we saw in Section 13.1, what many of us do at work these days is to ingest information, process it and pass it on to someone else. Intelligent machines are getting better at all of this than us. They are already better than us at image recognition; they are overtaking us in speech recognition, and they are catching up in natural language processing. Unlike us, they are improving fast: thanks to Moore's law (which is morphing, not dying), they get twice as good every 18 months or so.

To argue that technological unemployment cannot happen in the next few decades simply because it has not happened in the past is to commit a logical fallacy: the reverse Luddite fallacy.[126]

13.4.2 Inexhaustible Demand

Some people argue that humans will not become unemployable because there is an inexhaustible well of potential demand. In an blog post in June 2014, Marc Andreessen wrote 'This sort of thinking is textbook Luddism, relying on a "lump-of-labor" fallacy – the idea that there is a fixed amount of work to be done. The counter-argument to a finite supply of work comes from economist Milton Friedman – Human wants and needs are infinite, which means there is always more to do'.[127]

There may be plenty of demand, but if a machine can always provide the supply cheaper, better and faster, it will always be economically compelling to have it do the job instead of a human.

13.4.3 Icebergs

We saw in the last section that applying AI systems to what lawyers call the discovery, or disclosure process looked set to remove the need for junior lawyers, but instead it turned out that the machines brought a huge new class of work into the realm of possibility. Projects that would previously have been uneconomic were now feasible, and the junior lawyers are still needed to carry out the initial training. This has been called the paradox of automation, or the iceberg phenomenon: it was thought the junior lawyers were standing on thin ice, but it turned out instead that they were standing on top of a massive bulk of newly available work. Their position began to look secure again.

A similar phenomenon is likely to be observed with medical diagnosis. Again in Section 3.9, we saw that cheap attachments to smartphones will soon enable us all to be tested far more often and far more cheaply than at the moment. The AI resident on your smartphone – or servers it accesses in the cloud – will assess your blood pressure, blood glucose, your breath, your voice and many more indicators, and deliver instant verdicts in most cases. Initially at least, the upshot will be that we will have much better information about our medical condition, and we will need more doctors, not fewer, to explain the test results to us.

However, over time, the machines will continue to get faster and more effective, and work their way down the iceberg, so all those new jobs will disappear again.

13.4.4 The Human Touch

Some observers think that our salvation from machine intelligence automation lies in our very humanity. Our social skills and our ability to empathise and to care mean that we carry out tasks in a different way than machines do. Machines are by definition impersonal, the argument goes, and this renders them unsuitable for some types of job.

David Deming, a research fellow at the US National Bureau of Economic Research, believes we are already seeing the implications of this. In a report published in 2015, he claimed that the fastest growth in US employment since as long ago as 1980 has been in jobs requiring good social skills. Jobs requiring strong analytical abilities but no social skills have been in decline – with the implication that they are already being automated.[128]

Unfortunately, it isn't true that humans want to deal with other humans whenever possible. The first automatic deposit machine, the Bankograph, was installed in a bank in New York in 1960, but it was rejected by its intended customers. Its inventor, Luther Simjian, explained that 'The only people using the machines were prostitutes and gamblers who didn't want to deal with tellers face to face', and there were not enough of them to make the machines a worthwhile investment.[129] The first cash dispensing machine, or ATM, was installed in a bank in North London in June 1967. At first, people were hesitant to use it, but that changed when they realised they no longer had to queue for their cash, and they could access it when the banks were closed (which was most of the time, in those days). Very quickly, people showed a marked preference for the machine over the human bank teller.[130]

Nursing is an occupation long associated with caring people. Images of Florence Nightingale emoting as she nursed the wounded of the Crimean War are deeply ingrained in the profession's self-image. But there is growing evidence that robots make perfectly acceptable companions for sick people, and are sometimes preferable to their human equivalents. The Paro is a robotic seal developed for use in hospitals. Cute-looking, with big black eyes and covered in soft fur, it contains two 32-bit processors, three microphones, 12 tactile sensors, and it is animated by a system of silent motors. It recharges by sucking on a fake baby pacifier.

The Paro cost $15 million to develop; it distinguishes between individual humans, and repeats behaviours which appear to please them.[131] It has proved especially popular with patients suffering from dementia. As Shannon Vallor, a philosophy professor at Santa Clara University remarked, 'People have demonstrated a remarkable ability to transfer their psychological expectations of other people's thoughts, emotions, and feelings to robots'.[132]

Japan is the test-bed for the acceptability of robotic elder care. Thanks to a consistently low birth rate it has a greying population, and it has long been resistant to mass immigration as a solution for labour shortages. Furthermore its population is generally technophile. It is probably no coincidence that Softbank, one of Japan's largest companies, now owns several of the world's leading robotics companies, including Aldebaran and Boston Dynamics. The evidence so far is that the acceptability is high. A manager at a Tokyo home for the elderly says, 'A lot of people thought that elderly people would be scared or uncomfortable with robots, but they are actually very interested and interact naturally with them. They really enjoy talking to them and their motivation goes up when they use the rehabilitation robots, helping them to walk again more quickly'.[133]

So, humans are happy to interact with machines more often than we might intuitively expect. Furthermore, machines are much better at understanding humans than we might expect.

13.4.5 Robot Therapist

The US Army has a big problem with PTSD among veterans, not least because soldiers don't like to admit they have it. DARPA funds research at the University of Southern California to develop online therapy services, and the latest result is an online virtual therapist called Ellie.[134] She is proving to be better than human therapists at diagnosing PTSD.

There are two reasons for this. First, soldiers feel less embarrassed discussing their feelings with an entity they know will not judge them. In one test, 100 subjects were told that Ellie was controlled by a human, and another 100 were told that it was a robot. This second group displayed their feelings more openly, both verbally and in their expressions.[135]

Secondly, and perhaps more interesting, Ellie gleans most of its information about what is going on inside the soldier's head from his facial expressions rather than from what he says. When talking to a human therapist, the soldier may successfully 'sell' the idea that there is nothing wrong, because the human therapist listens closely to what he says, and may miss the subtle facial signals that contradict him. Counter-intuitively, people with depression smile just as frequently as happy people, but their smiles are shorter and more forced. Ellie is superb at catching this.[136]

Most people would probably agree with David Deming when he says that 'Reading the minds of others and reacting is [a skill that] has evolved in humans over thousands of years. Human interaction in the workplace involves team production, with workers playing off of each other's strengths and adapting flexibly to changing circumstances. Such non-routine interaction is at the heart of the human advantage over machines'. But we may soon have to rethink that.

It is far from clear that there could ever be enough jobs in the so-called caring professions to employ all the people who would in previous generations have been drivers, doctors, lawyers, management consultants and so on. Especially if machines are muscling into the caring professions too.

13.4.6 Made by Hand

Another way that people have suggested the human touch could preserve employment is that we will place a higher value on items manufactured by humans than on items manufactured by machines. It is hard to see much evidence of this in today's world outside some niche areas like handmade cakes.[137] Not many people today buy handmade radios or handmade cars.

There are four reasons why people might prefer products and services made by humans rather than machines: quality, loyalty, variation and status.

If humans produce a better product or provide a better service than machines, then other humans will buy from them. But the argument of this chapter is that certainly in many areas, and probably in most, machines will produce goods and services cheaper, better and faster.

Loyalty to our species might be a better defence. 'Buy handmade, save a human!' sounds like a plausible rallying-cry, or at least a marketing slogan. The past is not always a reliable guide to the exponential future, but it is a good place to start, and unfortunately it does not augur well for appeals to loyalty. In the late 1960s, Britain was feeling queasy as the Empire dissolved and Germany's economic power was returning. The 'I'm backing

Britain' campaign started in December 1967, trying to get British people to buy domestically manufactured products instead of imports. It fizzled out within a few months.[138]

Car manufacturing has long been symbolic of a nation's manufacturing virility. In the 1950s, Britain was the world's second-largest car manufacturer after the United States, but in the 1960s its designs and build quality fell behind, first, its European rivals, and then the Japanese. Despite repeated appeals to buy British, sales declined, and in 1975 the remaining national manufacturer, British Leyland, was nationalised. It never recovered, and Britain is now home to none of the major global car brands. (Fortunately, it has many innovative and thriving automotive design and component businesses working for foreign brands.)

Appealing to people to buy handmade items out of loyalty to one's species may not have a huge economic impact if machine-made items are better quality and much, much cheaper. And in a world of falling employment, most people are going to have to buy as efficiently as they can.

The third reason for buying from humans could be summarised by the phrase 'artisanal variation'. We like antiques because the patina of age gives them personality: each one is unique. The same goes for the original work of an artist, even if it isn't a Vermeer or a Rubens. But for most people, this is the preserve of luxury items, a few select pieces which we keep on display. Most of our possessions are mass-produced because they are much cheaper, disposable, and we can afford a better lifestyle that way.

We have seen this before, in the second half of the nineteenth century. With the Industrial Revolution in full swing, William Morris helped found the arts and crafts movement to produce handmade furnishings and decorations. His concern was to raise quality rather than to reduce unemployment, but in practice he ended up making expensive pieces which only the rich could afford.[139]

Finally, some people may choose to buy goods and services from humans rather than machines for reasons of status. But by definition, this could only ever amount to a niche activity, and would not save most of us from unemployability.

13.4.7 Entrepreneurs

If machines take over the jobs that are repetitive, humans will look to do things that require creativity, intuition, and to pursue counter-intuitive paths. One job title which fits that description is entrepreneur.

In my experience there are two types of entrepreneur. Both are resource-ful, determined and usually of above-average intelligence. The first and most common type is someone who works in an organisation which is doing something poorly. They notice this, and decide to offer a better ver-sion. They utilise essential skills and industry know-how acquired while working for the original organisation and simply improve incrementally on what was being provided there. These people are talented and hard-working, but they also had the good fortune to be in the right place at the right time to spot the opportunity. If they had not been in that position, they would have spent their careers working for other people, and because they are hard-working and bright they would probably have made a good fist of that.

The second type is destined to be an entrepreneur whatever circum-stances life drops them into. They will never be happy working for someone else. They envision themselves in a future world which looks impossible to anyone else, but they choose to believe it and by dint of sheer force of will they make that future a reality. They will walk through brick walls to make it happen, and will probably be bankrupt more than once. They are charming, astonishingly energetic, and often rather hard to be around. In the words of LinkedIn founder Reid Hoffman, they are people who will happily throw themselves off a cliff and assemble an aeroplane on the way down.[140]

Both types of entrepreneur are rare, and especially the second kind – which may be a good thing for the rest of us. In any case, this is probably not an occupation that is going to save large numbers of people from technological unemployment. The other thing to remember about entrepreneurship as a career is that most start-ups fail.

13.4.8 Centaurs

A computer first beat the best human at chess back in 1997. Deep Blue was one of the most powerful computers in the world when it beat Garry Kasparov; the match was close and the result was controversial. Today, a programme running on a laptop could beat any human.

But Kasparov claims that a very good human chess player teamed up with a powerful chess computer can beat a second chess computer playing on its own. Humans can undermine the game of a computer by throwing in some surprise moves which don't make much sense in the short term, or by deploying an intuitive strategy. Matches between humans working with computers are called advanced chess, or centaur chess. Kasparov himself initiated the first high-level centaur chess competition in Leon, in Spain, in 1998, and competitions have been held there regularly ever since. Tyler Cowen explores this form of chess extensively in his book, *Average is Over*.

Some people believe this phenomenon of humans teaming up with computers to form centaurs is a metaphor for how we can avoid most jobs being automated by machine intelligence. The computer will take care of those aspects of the job (or task) which are routine, logical and dull, and the human will be freed up to deploy her intuition and creativity. Engineers didn't become redundant just because computers replaced slide rules. Kevin Kelly, founder of *Wired* magazine, puts it more lyrically: machines are for answers; humans are for questions.[141]

The trouble is that the intuition and creativity which we humans bring to tasks and jobs is largely a matter of pattern recognition, and machines are getting better at this at an exponential rate. A doctor may be happy to delegate the routine diagnosis of a cold or a flu to a machine which can do it better than she can, if she gets to retain the more interesting and challenging diagnostic work. But what is to stop the machine overtaking the doctor in the more difficult cases as well? The lawyer is in the same boat: the tedious business of sifting through a haystack of documents looking for the needle of evidence is already being outsourced to machines. The more interesting and demanding task of devising a legal strategy is likely to follow suit.

Admittedly, there may need to be some level of human supervision of machine work until the machines acquire a degree of common sense. Before then, the blindly logical thought processes of a machine will not realise when a data glitch or a software bug has generated a bizarre conclusion which is unworkable or dangerous. But as we saw in Chapter 3: Exponential Improvement, the founding father of deep learning thinks that machines with common sense will appear in a decade or so. This does not mean that they will acquire consciousness, but merely that they will create internal models of the external world which will enable them to appreciate the impacts of glitches and bugs, just as we do.

Machines have already made considerable progress automating routine tasks, and indeed whole occupations where all the tasks are routine. As their performance improves they will increasingly take over tasks and jobs which are non-routine.

In response to a survey published in May 2016, the veteran AI researcher Nil Nilsson suggested laconically that before long, machines would be singing the song Irving Berlin composed for the 1946 Broadway musical *Annie Get Your Gun*. The lyric is 'Anything you can do, I can do better. I can do anything better than you'.[142]

13.4.9 The Magic Jobs Drawer

If machines are going to take a great many, perhaps most, of our existing jobs, can we create a host of new ones – perhaps whole new industries – to replace them? Those who think we can point out that many of the jobs we do today did not exist a 100 years ago. Our grandparents would not have understood what we mean by website builder, social media marketer, user experience designer, chief brand evangelist and so on. Surely, the argument goes, all these new technologies we have been talking about will throw up many new types of jobs that we cannot imagine today.

As the person probably most responsible for Google's self-driving cars, Sebastian Thrun is a man worth listening to on the subject. He is optimistic: 'With the advent of new technologies, we've always created new jobs. I don't know what these jobs will be, but I'm confident we will find them'.[143]

Unfortunately, past experience is (again) not as encouraging as you might think. Gerald Huff is a senior software engineer working in Silicon Valley, ground zero of the developments we are talking about. Nervous about the prospect of technological unemployment, he carried out a comparative analysis of US occupations in 1914 and 2014. Using data from the US Department of Labour,[144] he discovered that 80% of the 2014 occupations already existed in 1914. Furthermore, the numbers of people employed in the 20% of new occupations were modest, with only 10% of the working population engaged in them. The US economy is much bigger today than it was in 1914, and employs far more people, but the occupations are not new.

Of course, those of us who argue that it is different this time cannot rely on the historical precedent. It might be different this time in that vast swathes of new jobs *will* be created – including jobs for averagely skilled people, not just relatively high-skill jobs like social media marketing. But those who argue that we are falling for the Luddite fallacy cannot argue that history points to everybody getting new types of jobs which are more interesting and safer after a period of adjustment. It doesn't.

If we *were* to create a host of new jobs, what might they be? Maybe some of us will become dream wranglers, guiding each other towards fluency in lucid dreaming. Others may become emotion coaches, helping each other to overcome depression, anxiety and frustration. Maybe there will be jobs for which we have no words today, because the technology has not yet evolved to allow them to come into being.

It's not hard to imagine that virtual reality will create a lot of new jobs. If it is addictive as enthusiasts think it will be, many people will spend a great deal of their time – perhaps the majority of it – in VR worlds. In that case, there will be huge demand for new and better imaginary or simulated worlds to inhabit, and that means jobs.

But does it mean jobs for humans? Although the credit list for the latest superhero blockbuster stretches all the way around the block as it names everyone involved in rotoscoping and compositing the hyper-realistic armies of aliens, the latest computer-generated imagery (CGI) technology also makes it possible for two teenagers with a mobile phone to make a film which gains theatrical distribution. Their increasingly powerful software and hardware allows Hollywood directors to conjure visual worlds of such compelling complexity that their predecessors would rub their eyes in disbelief, but it also allows huge quantities of immersive content to be developed by skeleton crews. There will probably always be an elite of directors who are highly paid to push the boundaries of what can be imagined and what can be created, but software will do more and more of the heavy lifting in VR production.

Not for the first time, the games industry shows what is possible. A game called 'No Man's Sky' was announced in 2014 which conjures far more imaginary worlds than you could visit in a lifetime purely by the operation of algorithms and random number generators. You boldly go where no programmer or designer has gone before.[145]

The historical record should not make us confident that we will invent hordes of new jobs to replaces the ones taken over by machines. It might happen, but the firm belief that it will is little better than a blind belief that there is a 'magic job drawer' that we can open when we need to, and thousands of exciting new jobs will fly out.

13.4.10 Artists

After all these apparently gloomy prognostications, let's strike a more optimistic note. There is one profession which can probably never be automated until the arrival of an AGI which is also fully conscious. That profession is art, and to understand why, it is important to distinguish between art and creativity.

Creativity is the use of imagination to create something original. Imagination is the faculty of having original ideas, and there seems to be no reason why that requires a conscious mind to be at work. Creativity can simply be the act of combining two existing ideas (perhaps from different domains of expertise) in a novel way.

The eminent nineteenth-century chemist August Kekule solved the riddle of the molecular structure of benzene while day-dreaming, gazing into a fire.[146] True, he had spent a long time before that pondering the problem, but according to his own account, his conscious mind was definitely not at work when the creative spark ignited. You might argue that Kekule's subconscious was the originator of the insight, and that a subconscious can only exist where there is consciousness, but that seems to me an assertion that needs proving.

We saw in Chapter 2: The State of the Art, that machines can be creative. In mid-2015, Google researchers installed a feedback loop in an image recognition neural network, and the result was a series of fabulously hallucinogenic images.[147] To deny that they were creative is to distort the meaning of the word.

Art is something different. Admittedly, this is a personal definition, and perhaps not everyone would agree, but surely art involves the application of creativity to express something of personal importance to the artist. It might be beauty, an emotion or a profound insight into what it means to be human. (If that disqualifies a good deal of what is currently sold under the banner of art, then so be it – in fact, three cheers.)

To say something about your own experience clearly requires you to have had some experience, and that requires consciousness. Therefore, until a conscious AGI arrives, AIs can be creative but not artistic. This in turn means that while Donna Tartt and Kazuo Ishiguro are probably OK for a few decades; today's successful genre writers, who use stables of assistants to churn out several crime and romance novels each year, have chosen the timing of their careers expertly, and their publishers had better find something different to do.

In April 2016, researchers from Microsoft, a Dutch university and two Dutch art galleries created an AI which analysed the way Rembrandt painted. It identified enough of his techniques and mannerisms to enable it to produce paintings in exactly his style – better than any human forger could. They had it design a new picture in Rembrandt's style, of a subject Rembrandt had never worked on, and 3D printed it to capture the old master's technique in three dimensions. Because the machine recognises patterns better than humans can, it may well teach us interesting new things about the way Rembrandt created his masterpieces. But it is not producing art.[148]

13.4.11 Education

It is surprising how many smart people think that education is the answer to automation by machine intelligence. Microsoft CEO Satya Nadella said in January 2016: 'I feel the right emphasis is on skills, rather than worrying too much about the jobs [which] will be lost. We will have to spend the money to educate our people, not just children but also people mid-career so they can find new jobs'.[149]

Massive open online courses, or MOOCs, are promoted as the way we will all retrain for a new job each time a machine takes our old one. MOOCs are important, and along with flipped lessons, competency-based learning, and the use of big data, they will improve the quality of education, and make excellent learning opportunities available to all.

(With flipped lessons, students watch a video of a lecture for homework, and then put what they have been told into practice in the classroom. The teacher acts as coach and mentor, a more interactive role than lecturing. Competency-based learning requires students to have mastered a skill or a lesson before they move on to the next one; students within a class may progress at different speeds. Big data enables students and teachers to understand how well the learning process is going, and where extra support is needed.)

Exciting and powerful as these techniques are, they won't protect us from technological unemployment. We have seen that machines are increasingly capable of performing many of the tasks currently carried out by highly educated, highly paid people. The machines aren't just coming for the jobs of bricklayers; they're coming for the jobs of surgeons and lawyers too.

There is a very important postscript to these remarks about education. If we make it through successfully to the new world in which many or most of us are permanently and irrevocably unemployed, then education will be more important, not less. We will need good education to take advantage of our leisured lives, even more than we did to survive our working lives. But the education we will need will be *vacational*, not *vocational*.

13.5 CONCLUSION: YES, IT'S DIFFERENT THIS TIME

In Chapter 2: The State of the Art, we considered the state of the art of AI, and in Chapter 3: Exponential Improvement, we saw how dramatic the impact of exponential growth can be. (No apologies for repeating that point: it is critical.) Then in Section 4.5 we reviewed the likely evolution of technologies which are associated with AI, and caused or enabled by it.

In Chapter 12: The History of Automation, we saw that previous rounds of automation in the Industrial Revolution did not cause unemployment of humans in the long term – although the long term is fairly long in this context: the Engels pause that we came across in Section 12.4 lasted at least a generation (a quarter-century). The horses were not so lucky.

In Sections 13.2 and 13.3, we discussed how various occupations could be automated, starting with the poster child of driving vehicles, and concluding with the privileged elites in the professions and high finance.

Finally, in Section 13.4 we reviewed the arguments of sceptics, who believe that technological unemployment will not happen.

It's time to answer the question: is it really different this time? Will machine intelligence automate most human jobs within the next few decades and leave a large minority of people – perhaps a majority – unable to gain paid employment?

It seems to me that you have to accept that this proposition is at least possible if you admit the following three premises:

1. It is possible to automate the cognitive and manual tasks that we carry out to do our jobs.

2. Machine intelligence is approaching or overtaking our ability to ingest, process and pass on data presented in visual form and in natural language.

3. Machine intelligence is improving at an exponential rate. This rate may or may not slow a little in the coming years, but it will continue to be very fast.

No doubt it is still possible to reject one or more of these premises, but for me, the evidence assembled in this chapter makes that hard. I can't escape the conclusion that widespread and lasting technological unemployment may well be coming within a generation or two.

It hasn't started yet: the United States and the United Kingdom are pretty close to full employment (although some would argue there is plenty of joblessness disguised as self-employment). When it does arrive, it will probably be the way Hemingway described going bankrupt in his 1926 novel *The Sun Also Rises*: 'How did you go bankrupt?' Bill asked. 'Two ways', Mike said. 'Gradually and then suddenly'. Companies dislike sacking lots of people: it is bad for morale and bad PR. Insofar as they can, they will redeploy people whose jobs have been automated and cut down on the general recruiting. This will disguise the process for a while, but people will notice it is getting harder to find new jobs. Whenever there is a crisis, companies will take the opportunity to let some people go. Eventually, they won't be able to hide what is happening, and the trickle will become a flow.

Somewhere in this process there is likely to be a panic. We'll get to that in the next chapter.

13.5.1 Confidence

Please note that I don't claim to know for certain that technological unemployment will happen. I don't have a crystal ball. But it seems likely, and it certainly seems sufficiently plausible, that we should make plans for what to do if it does happen.

I am often surprised at the vehemence of people who think it won't happen. For instance, in May 2017, Eric Schmidt proudly proclaimed himself a 'job elimination denier', saying that the transformation we are going through will be no different to previous economic revolutions.[150] To me, this approach seems dangerous.

If technological unemployment never arrives and we have spent a modest amount of resources contingency planning for it, as I will suggest in Section 17.2, then we will have lost very little. On the other hand, it if does arrive and we have done nothing, the consequences could be grave. We should avoid that.

13.5.2 Optimism

People who think technological unemployment is probably coming are often described as pessimists. Some of them are: they are building secure hiding places in remote locations because they think social collapse is coming. But this is not the only possible response.

I am an optimist. I think a world in which machines do most of the jobs could be a world in which humans get on with the important parts of life: playing, having fun, learning, exploring, being creative, playing sports, socialising, inventing things.

Surely the pessimists are the people who insist that all humans must continue doing jobs forever. Many people do enjoy their jobs, but most don't. A Gallup poll in 142 countries in 2013 found that only one in eight people were positively engaged in their work.[151] For most people, their job is simply the way they pay for food, clothes and housing for themselves and their families.

Financial Times columnist Martin Wolf urged that we should 'enslave the robots and free the poor',[152] and who would not welcome such an outcome?

Unfortunately, a smooth transition to a leisure society may well not be the default outcome, and in the next chapter we will review some of the challenges we face in getting there. But let's pause for breath for a moment, and beguile ourselves with Richard Brautigan's poetic wish, written back in 1967.

> I like to think
> (it has to be!)
> of a cybernetic ecology
> where we are free of our labors
> and joined back to nature,
> returned to our mammal
> brothers and sisters, and all watched over
> by machines of loving grace.[153]

NOTES

1. Assuming the work is happening on Earth. Wikipedia offers a more general but less euphonious definition: 'Work is the product of the force applied and the displacement of the point where the force is applied in the direction of the force'.

2. http://www.mckinsey.com/insights/business_technology/four_fundamentals_of_workplace_automation.

3. 2013 data: http://www.ons.gov.uk/ons/dcp171778_315661.pdf.

4. https://www.invespcro.com/blog/global-online-retail-spending-statistics-and-trends/.

5. This has given rise to the term 'subtractive manufacturing' for the traditional forms of manufacturing. This method of naming is rather splendidly called a retronym.

6. https://en.wikipedia.org/wiki/Fax.

7. http://money.cnn.com/2017/05/24/news/economy/gig-economy-intuit/index.html.

8. WHO, *Global Status Report on Road Safety 2013: Supporting a Decade of Action.*

9. http://www.japantimes.co.jp/news/2015/11/15/business/tech/human-drivers-biggest-threat-developing-self-driving-cars/#.Vo7D5fmLRD8.

10. http://www.theatlantic.com/business/archive/2013/02/the-american-commuter-spends-38-hours-a-year-stuck-in-traffic/272905/.

11. http://www.reinventingparking.org/2013/02/cars-are-parked-95-of-time-lets-check.html.

12. http://www.etymonline.com/index.php?term=autocar.

13. http://www.digitaltrends.com/cars/audi-autonomous-car-prototype-starts-550-mile-trip-to-ces/.

14. http://www.reuters.com/investigates/special-report/autos-driverless/.

15. http://www.wired.com/2015/04/delphi-autonomous-car-cross-country/.

16. http://www.reuters.com/investigates/special-report/autos-driverless/.

17. https://www.theatlantic.com/technology/archive/2017/08/inside-waymos-secret-testing-and-simulation-facilities/537648/.

18. http://techcrunch.com/2015/12/22/a-new-system-lets-self-driving-cars-learn-streets-on-the-fly/.

19. http://www.bloomberg.com/news/articles/2015-12-16/google-said-to-make-driverless-cars-an-alphabet-company-in-2016.

20. https://spectrum.ieee.org/cars-that-think/transportation/self-driving/google-has-spent-over-11-billion-on-selfdriving-tech.

21. https://mondaynote.com/the-sd-cars-gold-rush-winner-bee2b5ae7b21.

22. http://venturebeat.com/2016/01/10/elon-musk-youll-be-able-to-summon-your-tesla-from-anywhere-in-2018/.

23. https://www.washingtonpost.com/news/the-switch/wp/2016/01/11/elon-musk-says-teslas-autopilot-is-already-probably-better-than-human-drivers/.

24. http://electrek.co/2016/04/24/tesla-autopilot-probability-accident/.

25. https://techcrunch.com/2017/09/21/baidu-announces-1-5b-fund-to-back-self-driving-car-startups/?utm_medium=TCnewsletter.
26. http://www.pcmag.com/article2/0,2817,2370598,00.asp.
27. http://www.nytimes.com/2015/11/06/technology/toyota-silicon-valley-artificial-intelligence-research-center.html?_r=0.
28. https://www.yahoo.com/autos/google-pairs-with-ford-to-1326344237400118.html.
29. http://uk.businessinsider.com/bmw-says-cars-with-artificial-intelligence-are-already-here-2016-1?r=US&IR=T.
30. http://www.bbc.co.uk/news/technology-35280632.
31. http://www.zdnet.com/article/ford-self-driving-cars-are-five-years-away-from-changing-the-world/.
32. http://cleantechnica.com/2015/10/12/autonomous-buses-being-tested-in-greek-city-of-trikala/.
33. http://www.businessinsider.com/toyota-gill-pratt-unveils-self-driving-plans-concept-car-at-ces-2017-1.
34. http://recode.net/2015/03/17/google-self-driving-car-chief-wants-tech-on-the-market-within-five-years/.
35. https://www.recode.net/2017/9/8/16278566/transcript-self-driving-car-engineer-chris-urmson-recode-decode.
36. https://www.cnbc.com/2017/05/23/ford-panicking-self-driving-cars-because-alphabet-google-is-way-ahead.html.
37. https://arstechnica.com/cars/2017/10/report-waymo-aiming-to-launch-commercial-driverless-service-this-year/.
38. https://www.navigantresearch.com/research/navigant-research-leaderboard-report-automated-driving.
39. https://media.ford.com/content/fordmedia/fna/us/en/news/2016/08/16/ford-targets-fully-autonomous-vehicle-for-ride-sharing-in-2021.html.
40. https://www.cnbc.com/2017/09/25/gm-developing-autonomous-vehicles-at-a-fast-pace-deutsche-bank-says.html.
41. https://electrek.co/2017/09/06/tesla-semi-all-electric-truck-biggest-catalys/.
42. http://www.wired.com/2015/12/californias-new-self-driving-car-rules-are-great-for-texas/.
43. http://www.reuters.com/investigates/special-report/autos-driverless/.
44. It has been suggested that electric cars should make noises so that people don't step off the pavement in front of them. A friend tells me he would like his to make a noise like two coconuts being banged together, in homage to the scene in Monty Python and the Holy Grail where King Arthur, unable to afford a horse, has a camp follower fake the noise of one with coconuts.
45. http://www.wsj.com/articles/SB100014240531119034809045765122509156299460.
46. http://fortune.com/2014/05/04/6-things-i-learned-at-buffetts-annual-meeting/.
47. http://www.thenewspaper.com/news/43/4341.asp.
48. http://www.alltrucking.com/faq/truck-drivers-in-the-usa/.

49. http://www.bls.gov/ooh/transportation-and-material-moving/bus-drivers.htm.
50. http://www.bls.gov/ooh/transportation-and-material-moving/taxi-drivers-and-chauffeurs.htm.
51. http://www.joc.com/trucking-logistics/truckload-freight/driver-wage-hikes-could-raise-truckload-pricing-12-18-percent_20150325.html.
52. http://bbc.in/2uTK2PD
53. http://www.abc.net.au/news/2015-10-18/rio-tinto-opens-worlds-first-automated-mine/6863814.
54. http://www.mining.com/why-western-australia-became-the-center-of-mine-automation/.
55. http://www.npr.org/sections/money/2015/02/05/382664837/map-the-most-common-job-in-every-state.
56. http://www.huffingtonpost.com/entry/pittsburgh-uber-self-driving-cars_us_59caa91ae4b0d0b254c4fcdf?ncid=inblnkushpmg00000009.
57. The Programme was established in January 2015 with funding from Citibank, one of the largest financial institutions in the world. The Oxford Martin School was set up as part of Oxford University in 2005, as an institution dedicated to understanding the threats and opportunities facing humanity in the twenty-first century. It is named after James Martin, a writer, consultant and entrepreneur, who founded the school with the largest donation ever made to the university – which was no mean feat given that Oxford was founded 1,000 years ago, and is the oldest university in the world (after Bologna in Italy).
58. http://www.oxfordmartin.ox.ac.uk/downloads/academic/The_Future_of_Employment.pdf.
59. https://www.ndtv.com/world-news/face-scans-robot-baggage-handlers-airports-of-the-future-1742507.
60. https://www.technologyreview.com/s/608811/drones-and-robots-are-taking-over-industrial-inspection/.
61. https://en.wikipedia.org/wiki/Horn_%26_Hardart#Automated_food.
62. http://www.computerworld.com/article/2837810/automation-arrives-at-restaurants-but-dont-blame-rising-minimum-wages.html.
63. http://blogs.forrester.com/andy_hoar/15-04-14-death_of_a_b2b_salesman.
64. http://www.nuance.com/for-business/customer-service-solutions/nina/index.htm.
65. http://www.zdnet.com/article/swedbank-humanises-customer-service-with-artificial-intelligence-platform/.
66. https://spectrum.ieee.org/automaton/robotics/industrial-robots/autonomous-robots-plant-tend-and-harvest-entire-crop-of-barley.
67. https://www.fastcompany.com/40473583/this-strawberry-picking-robot-gently-picks-the-ripest-berries-with-its-robo-hand?partner=recode.
68. https://www.technologyreview.com/s/601215/china-is-building-a-robot-army-of-model-workers/.
69. https://www.cnbc.com/2017/09/11/new-york-times-digital-as-amazon-pushes-forward-with-robots-workers-find-new-roles.html.

70. https://www.invespcro.com/blog/global-online-retail-spending-statistics-and-trends/.

71. https://www.wired.com/story/grasping-robots-compete-to-rule-amazons-warehouses/.

72. http://www.theguardian.com/technology/2014/sep/12/artificial-intelligence-data-journalism-media.

73. http://www.arria.com/

74. https://www.youtube.com/watch?v=HXKDnqM9Ulw.

75. http://www.chinadaily.com.cn/china/2015-12/24/content_22794242.htm.

76. http://persado.com/.

77. http://markets.businessinsider.com/news/stocks/artificial-intelligence-could-make-brands-obsolete-2017-9-1002515382.

78. http://www.techtimes.com/articles/127526/20160126/ai-politics-how-an-artificial-intelligence-algorithm-can-write-political-speeches.htm.

79. http://www.ravn.co.uk/.

80. http://www.legalweek.com/legal-week/sponsored/2434504/is-artificial-intelligence-the-key-to-unlocking-innovation-in-your-law-firm.

81. http://linkis.com/www.theatlantic.com/SoE5e.

82. http://www.legalfutures.co.uk/latest-news/come-americans-legalzoom-gains-abs-licence.

83. https://www.fairdocument.com/.

84. http://msutoday.msu.edu/news/2014/using-data-to-predict-supreme-courts-decisions/.

85. http://uk.businessinsider.com/robots-may-make-legal-workers-obsolete-2015-8.

86. http://www.telegraph.co.uk/news/2017/08/04/legal-robots-deployed-china-help-decide-thousands-cases/.

87. http://www.kurzweilai.net/machine-learning-rivals-human-skills-in-cancer-detection.

88. http://uk.businessinsider.com/deepmind-cofounders-invest-in-babylon-health-2016-1.

89. http://singularityhub.com/2016/01/18/digital-diagnosis-intelligent-machines-do-a-better-job-than-humans/?utm_content=bufferb9e5d&utm_medium=social&utm_source=twitter.com&utm_campaign=buffer.

90. http://forbesindia.com/article/hidden-gems/thyrocare-technologies-testing-new-waters-in-medical-diagnostics/41051/1.

91. http://www.ucsf.edu/news/2011/03/9510/new-ucsf-robotic-pharmacy-aims-improve-patient-safety.

92. http://www.qmed.com/news/ibms-watson-could-diagnose-cancer-better-doctors.

93. http://www.ft.com/cms/s/2/dced8150-b300-11e5-8358-9a82b43f6b2f.html#axzz3xL3RoRdy.

94. http://www.ft.com/cms/s/2/dced8150-b300-11e5-8358-9a82b43f6b2f.html#axzz3xL3RoRdy.

95. http://www.theverge.com/2016/3/10/11192774/demis-hassabis-interview-alphago-google-deepmind-ai.

96. http://qz.com/567658/searching-for-eureka-ibms-path-back-to-greatness-and-how-it-could-change-the-world/.
97. http://www.forbes.com/sites/peterhigh/2016/01/18/ibm-watson-head-mike-rhodin-on-the-future-of-artificial-intelligence/#24204aab3e2922228b9c30cc.
98. http://www.dotmed.com/news/story/29020.
99. http://www.mckinsey.com/global-themes/future-of-organizations-and-work/the-evolution-of-employment-and-skills-in-the-age-of-ai.
100. http://www.wsj.com/articles/SB100014240527023039839045790932525738 14132.
101. http://www.outpatientsurgery.net/outpatient-surgery-news-and-trends/general-surgical-news-and-reports/ethicon-pulling-sedasys-anesthesia-system--03-10-16.
102. http://www.wired.co.uk/news/archive/2016-05/05/autonomous-robot-surgeon.
103. http://www.scmp.com/news/china/article/2112197/chinese-robot-dentist-first-fit-implants-patients-mouth-without-any-human?utm_source=t.co&utm_medium=referral.
104. http://www.thelancet.com/journals/lanonc/article/PIIS1470-2045(17)30572-7/fulltext?elsca1=tlpr
105. https://www.edsurge.com/news/2016-04-18-gradescope-raises-2-6m-to-apply-artificial-intelligence-to-grading-exams.
106. http://www.wsj.com/articles/if-your-teacher-sounds-like-a-robot-you-might-be-on-to-something-1462546621
107. https://www.sigfig.com/site/#/home.
108. http://www.nytimes.com/2016/01/23/your-money/robo-advisers-for-investors-are-not-one-size-fits-all.html?_r=0.
109. https://itunes.apple.com/gb/podcast/exchanges-at-goldman-sachs/id948913991?mt=2&i=361020299.
110. https://www.theguardian.com/business/2017/sep/06/deutsche-bank-boss-says-big-number-of-staff-will-lose-jobs-to-automation?CMP=share_btn_tw
111. https://www.thenational.ae/business/mashreq-to-shed-10-per-cent-of-headcount-in-next-12-months-as-artificial-intelligence-spending-pays-off-1.628437.
112. http://www.businessinsider.com/r-technology-could-help-ubs-cut-work-force-by-30-percent-ceo-in-magazine-2017-10?IR=T
113. http://www.bloomberg.com/news/articles/2015-02-27/bridgewater-is-said-to-start-artificial-intelligence-team.
114. http://www.wired.com/2016/01/the-rise-of-the-artificially-intelligent-hedge-fund/.
115. https://next.ft.com/content/c31f8f44-033b-11e6-af1d-c47326021344 (Paywall).
116. https://www.bloomberg.com/news/features/2017-09-27/the-massive-hedge-fund-betting-on-ai.
117. http://www.ft.com/cms/s/0/5eb91614-bee5-11e5-846f-79b0e3d20eaf.html#axzz3zEmSvuZs.

118. https://www.bloomberg.com/news/features/2017-09-27/the-massive-hedge-fund-betting-on-ai.
119. https://www.ft.com/content/16b8ffb6-7161-11e7-aca6-c6bd07df1a3c.
120. http://uk.businessinsider.com/high-salary-jobs-will-be-automated-2016-3
121. http://www.fiercefinanceit.com/story/will-regulatory-compliance-drive-artificial-intelligence-adoption/2016-01-05.
122. http://www.liverpoolecho.co.uk/news/business/liverpool-fc-sponsor-standard-chartered-11104215.
123. http://www.cnbc.com/2015/12/30/artificial-intelligence-making-some-bosses-nervous-study.html.
124. https://www.wired.com/story/googles-learning-software-learns-to-write-learning-software/?mbid=social_twitter_onsiteshare.
125. http://www.eastoftheweb.com/short-stories/UBooks/BoyCri.shtml.
126. This useful phrase was originally coined by my friends at the Singularity Bros podcast, http://singularitybros.com/.
127. http://blog.pmarca.com/2014/06/13/this-is-probably-a-good-time-to-say-that-i-dont-believe-robots-will-eat-all-the-jobs/.
128. http://uk.businessinsider.com/social-skills-becoming-more-important-as-robots-enter-workforce-2015-12.
129. http://www.history.com/topics/inventions/automated-teller-machines.
130. http://www.theatlantic.com/technology/archive/2015/03/a-brief-history-of-the-atm/388547/.
131. http://www.wsj.com/articles/SB10001424052748704463504575301051844937276.
132. http://kalw.org/post/robotic-seals-comfort-dementia-patients-raise-ethical-concerns#stream/0.
133. http://www.scmp.com/week-asia/business/article/2104809/why-japan-will-profit-most-artificial-intelligence.
134. http://viterbi.usc.edu/news/news/2013/a-virtual-therapist.htm.
135. http://observer.com/2014/08/study-people-are-more-likely-to-open-up-to-a-talking-computer-than-a-human-therapist/.
136. http://mindthehorizon.com/2015/09/21/avatar-virtual-reality-mental-health-tech/.
137. http://www.handmadecake.co.uk/.
138. http://www.bbc.co.uk/news/magazine-15551818.
139. http://www.oxforddnb.com/view/article/19322.
140. http://www.inc.com/john-brandon/22-inspiring-quotes-from-famous-entrepreneurs.html.
141. https://www.edge.org/conversation/kevin_kelly-the-technium.
142. https://www.singularityweblog.com/techemergence-surveys-experts-on-ai-risks/.
143. http://www.ft.com/cms/s/2/c5cf07c4-bf8e-11e5-846f-79b0e3d20eaf.html#axzz3yLGlrrlJ.
144. http://www.bls.gov/cps/cpsaat11.htm.
145. https://en.wikipedia.org/wiki/No_Man%27s_Sky.
146. http://www.uh.edu/engines/epi265.htm.

147. http://googleresearch.blogspot.co.uk/2015/06/inceptionism-going-deeper-into-neural.html.
148. http://www.bbc.co.uk/news/technology-35977315.
149. http://www.ft.com/cms/s/2/c5cf07c4-bf8e-11e5-846f-79b0e3d20eaf.html#axzz3yLGlrr1J.
150. http://uk.businessinsider.com/googles-eric-schmidt-im-a-job-elimination-denier-on-the-risk-of-robots-stealing-jobs-2017-5.
151. http://www.gallup.com/poll/165269/worldwide-employees-engaged-work.aspx.
152. http://www.ft.com/cms/s/0/dfe218d6-9038-11e3-a776-00144feab7de.html#axzz3yUOe9Hkp.
153. http://www.brautigan.net/machines.html.

CHAPTER 14

The Challenges

The point of this book so far has been to persuade you that within a few decades, it is likely that many people will be rendered unemployable by machine intelligence. If I have not wholly succeeded in that aim, then I hope you are at least prepared to accept that the possibility is serious enough that we should be thinking about the implications, and what to do about it if it happens.

If I haven't even got you that far, then you're probably about to put this book down. If so, don't throw it away – you might want to come back to it when self-driving vehicles start to make serious impacts on the employment data.

If I have made the case successfully – or if you were persuaded before we started – then welcome to the next stage of the journey. If we are en route to a leisure society, what do we have to do to make the destination wonderful and the transition smooth?

There will be challenges. I anticipate six: meaning, economic contraction, income, allocation, cohesion and panic. Let's take a look at each of them in turn.

14.1 MEANING

14.1.1 The Meaning of Life

… is 42, of course.[1]

OK, now we've got that out of the way, would you agree with the statement that people's lives need to have meaning in order for them to feel fulfilled, satisfied and happy? It's certainly true for me, and I'm pretty sure it's true for most of the people I know. It is probably also true of you, or you wouldn't be reading this book. The initial reaction of many people when they first take seriously the possibility of widespread lasting technological unemployment is, 'how will we fill our days? How will we find meaning in our lives without work?'

I have met people who claimed to be pure hedonists – interested only in immediate pleasure. Some of them may even have been telling the truth. But most of us get bored if we feel our lives have no meaning. And not just bored in the sense that you get bored in a queue at a supermarket check-out, but profoundly restless and frustrated. To avoid this feeling we make deep emotional investments in ideas and institutions like family, friend-ships, work, loyalty to tribes, nations and causes. Deprived of these things, we feel lost and alienated.

Perhaps the most famous quote attributed to the fourth century BC Greek philosopher Socrates is that the unexamined life is not worth liv-ing. It is a remarkably strong statement. Why not just say that an unexam-ined life – a life without philosophy, in other words – is less good than an examined one? Is an unexamined life really worse than death? He made the statement at his trial, when the outcome was a choice between exile and suicide (he chose suicide), so perhaps he was under stress and being hyperbolic. But the claim is usually taken at face value, and perhaps he meant it literally.

It is also an elitist statement. Many people are too preoccupied with making a living, raising a family, escaping drug addiction or whatever immediate challenge they face to indulge in the luxury of philosophical discourse. Are their lives not worth living? You could argue that Socrates and his fellow ancient Athenians had slaves to take care of the menial stuff, but we have labour-saving devices instead, so that's no excuse.

Of course, the question of what constitutes a good life, a worthwhile life, a life with meaning is a vexed one, with no simple answers, and probably no single answer. The philosopher John Danaher distinguishes between subjective accounts, which involve *feeling* worthwhile, and objective accounts, which involve helping to *make or do* something worthwhile.[2]

Despite not knowing (or at least not agreeing) what a meaningful life is, and despite not spending all that much time in the average day thinking about it, most of us believe we need it. And many of us find it in work. So it's going to be a problem if we stop working.

Or is it?

14.1.2 Meaning and Work

Simon Sinek has made a name for himself with books that propound a simple but important truth: if you have a clear purpose which inspires others, you can achieve great things. His best-known saying is 'Working hard for something we don't care about is called stress; working hard for something we love is called passion'.

You could be forgiven for thinking that a law was passed a few years ago in the United States requiring business leaders – and people who want to be business leaders – to talk about their passion for their business. But most people don't feel passionate about their work, even if they pretend they do. In fact, many people are positively alienated by their jobs. They find them meaningless and boring.

Yet even these people usually define themselves by what they do for a living. If you ask someone at a party what they do they are likely to reply that they are an accountant, a taxi driver or an electrician. They are less likely to say that they are the coach of their child's football club, or a cinema-goer, or a reader. No doubt this is partly due to the amount of time that our jobs absorb – but then again we don't define ourselves as sleepers. It also has to do with work being the activity that provides our income, which is why home-based parents often feel sheepish about naming that as their work. (They shouldn't: parenting is one of the hardest but most rewarding jobs I've ever done!)

So jobs helps define us, and they give many of us purpose. They even give some of us meaning. How damaging would it be if we lost them? Unemployed people often struggle with depression, but they are experiencing it in the context of a society where it seems that everyone else has a job. They are also on a lower income than the employed people around them. How bad would it be if everyone else was also unemployed and receiving a decent income?

Fortunately, there are a couple of places we can look for an answer to that question.

14.1.3 The Rich and the Retired

The agricultural revolution, around 12,000 years ago, created sustainable surpluses of food and other basic resources. This enabled a class of people to stop doing the work that pretty much all humans had done since our arrival on the planet, which was foraging and hunting for food. They became tribal leaders, kings, warriors, priests, traders and so on. Sometimes they spent as much time on these activities as the people who continued to forage and hunt, but sometimes they took time off – deliberately or by happenstance – and engaged in lives of leisure.

In Europe, these people became known as aristocrats, from a Greek word meaning the best – originally in a military sense and then a political one. Some aristocrats did jobs: they ran agricultural concerns, they got involved in politics, and in some countries, they ran empires. Occasionally, they became men (and more rarely, women) of what we now call science. Famously, they disdained trade and commerce, regarding those activities as the preserve of the class below them, the middle class.

Many aristocrats did not work – including almost all the female ones. They led lives of leisure. As young men (and in a few cases, young women) they toured classical Greek and Roman sites in the Mediterranean countries. Returning home, they mostly socialised. Their lives revolved around balls, hunts and visits to their local peers, interspersed with the glamour and tragedy of war, if that was their inclination. This lifestyle was chronicled in the novel, an art form which first acquired its current realistic form in the early eighteenth century.[3]

The lives depicted by Jane Austen and her contemporaries may seem tame to modern readers, who have experienced international travel and expect simultaneous global communications. But they were agreeable lives compared with what their poorer contemporaries had to put up with. Addictions to gambling and drink were a hazard, and of course a minority of this pampered class destroyed themselves and their families with these vices. But this was unusual, and by and large most eighteenth- and nineteenth-century European aristocrats seem to have passed their lives without great concern about their lack of meaning. Whether these lives were worthwhile or not, whether or not they had meaning, is probably not for us to judge, but there is no evidence of widespread existential angst among the nobility.

In fact, it is these privileged people who made most of the advances in human thought and art in previous centuries, precisely because they did not need to work for a living, or eke out an existence as subsistence farmers. If they did not produce the memorable work themselves, they often sponsored it by employing talented artisans. So it seems there is much to be said for the ability to be idle.[4]

The other group we can look to for evidence about the effects of joblessness are retired people living on good incomes. The conventional wisdom used to be that growing old was an almost unmitigated disaster: 'Old age ain't no place for sissies', as Bette Davies said,[5] although it's obviously better than the only alternative currently available. But starting in the 1990s, researchers began questioning this perception, and found instead that the progress of happiness throughout life is U-shaped. We are at our happiest and most fulfilled when young, we become stressed and discontented in our prime and middle age, and we are happier and more relaxed again when older, despite the onset of physical disabilities and limitations.[6] This pattern has been observed across a wide range of societies, and over a substantial period of time.

There are probably numerous causes of this effect, including escape from the responsibility for looking after children, and the acquisition of wisdom, including an acceptance of what life has thrown at us. But the absence of jobs plays a major role in the lives of the retired. Even if it is not causing the up-tick in happiness, it is at least not preventing it.

Retiring in penury is no fun at all. But if you spend a few days in the towns and villages where better-off people retire – perhaps on defined benefit, final salary pensions – you will see very busy communities of people organising festivals and dinner parties, shuttling between games of bridge and lessons at the university of the third age. If you ask them how they spend their time, they will tell you they have no idea how they used to fit in their day jobs.

14.1.4 Virtually Happy

Thus far in human history we have had to find our meaning within the constraints of the three-dimensional world we live in, or in our imaginations. Technology is poised to open up a whole new space for us to explore together – the world of virtual reality (VR). We don't yet know how we will react to this new universe, how we will behave in it and what it will mean to us. We can be pretty confident that it will have a big impact.

Diaspora, Greg Egan's novel of the far future, features an environment called the 'Truth Mines'. It is a physical representation of mathematical theorems (albeit in VR) which can seemingly be explored forever without exhausting all the discoveries that can be made. The ability to create virtual worlds that are so convincing to our brains that we almost lose the understanding that they are artificial may well allow us to expand enormously the space within which we find happiness and meaning.

14.1.5 Helping Us Adjust

As we saw previously, only one in eight people around the world report being positively engaged in their work. But for the many people whose lives are given purpose – if not meaning – by their jobs, there may be some help required in adjusting to a jobless future. Many will need a little help to start, and then they will be fine. Others will continue to need periodic doses of support. Some will stumble badly, and will require extensive and extended assistance. We may have to do a lot of research and experimentation to get this right. But the examples of the retired and the aristocrats suggest to me that loss of meaning will not be one of the biggest problems that widespread technological unemployment will create.

14.2 ECONOMIC CONTRACTION

American union boss Walter Reuther recounts a story about a visit he made in the 1950s to a Ford manufacturing plant, where he saw an impressive array of robots assembling cars. The Ford executive who was showing him round asked how Reuther thought he would get the robots to pay union membership fees. Reuther replied that the bigger question was how the robots would buy cars. (The story is usually told with Henry Ford II playing the role of the company executive but it almost certainly wasn't.[7])

The basic economic problem which this story is supposed to illustrate is that if nobody is earning any money, then nobody can buy anything, and even those who do have money and resources can't sell anything. The economy grinds to a halt and everybody starves.

Of course, life is never as black-and-white as that. Economies don't go overnight from functioning tolerably well to complete collapse. Even catastrophic decline is less like falling off a cliff and more like tumbling down a slope, with pauses along the way as you hit ledges. But obviously, severe economic contraction is grim, and to be avoided if at all possible.

If and when machine intelligence renders more and more people unemployable, then other things being equal, the purchasing power previously exercised by those people will dry up. Their productive output will not be lost – it will just be provided by machines instead of humans. As demand falls but supply remains stable, prices will fall. At first, the falling prices may not be too much of a problem for firms and their owners, as the machines will be more efficient than the humans they replaced, and increasingly so, as they continue to improve at an exponential rate. But as more and more people become unemployed, the consequent fall in demand will overtake the price reductions enabled by greater efficiency. Economic contraction is pretty much inevitable, and it will get so serious that something will have to be done.

But before policy makers are forced to take action to tackle economic contraction, they will be faced by a much more serious problem: what to do about all those people who no longer have a source of income? Tackling this successfully will also solve the problem of economic contraction, so we can move right along.

14.3 INCOME

At the height of the Great Depression in the early 1930s, unemployment reached 25% of the working-age population.[8] Social security arrangements were primitive then, and developed societies were much poorer than they are today, so that level of joblessness was much harder on people than it is today, when parts of Europe have returned to similar levels overall,[9] with youth unemployment hitting 50% in some places.[10]

The worst levels of unemployment in developed countries today are found in Mediterranean countries like Greece and Spain, where family networks remain strong enough that sons and daughters can be supported for months or even years by fathers and mothers – and vice versa. There are escape valves, too, for the social pressure created by the situation. Economies further north are struggling less, and can absorb the energies and ambitions of many of the unemployed young people from the south.

When self-driving vehicles and other forms of automation render people of all classes unemployed right across the developed world, these safety nets will no longer be available. Articulate, well-connected and forceful middle-class professionals will be standing alongside professional drivers and factory workers, demanding that the state do something to protect them and their families.

14.3.1 Universal Basic Income

If and when societies reach the point where we have to admit that a significant proportion of the population will never work again – through no fault of their own – a mechanism will have to be found to keep those people alive. An answer which has become increasingly popular in the last few years is a universal basic income (UBI), available to all without condition; a living wage which is paid to all citizens simply because they are citizens.

Probably the longest-standing organisation advocating UBI is the Basic Income Earth Network (BIEN). BIEN was formed as long ago as 1986, and 'Earth' replaced 'European' in its name in 2004. BIEN defines UBI as 'an income unconditionally granted to all on an individual basis, without means test or work requirement'. UBI has also been called unconditional basic income, basic income, basic income guarantee (BIG), guaranteed annual income and citizen's income.

Proponents have argued for various levels of UBI, but in general they choose a level at or around the poverty level in the country of operation. This is partly because they understand that a higher level would be (even more) unaffordable, and partly to ward off criticisms that UBI would make people lazy and unproductive.

The benefits claimed for UBI address issues which concern both the political left and right. Left-wing proponents see it as a mechanism to eradicate poverty and redress what they view as growing inequality within societies. They sometimes argue that it tackles the alleged gender pay gap and redistributes income away from capital and towards labour. It has also been held out as a partial solution to the alleged generational theft whereby relatively wealthy pensioners are receiving income generated by taxes on young workers who have no assets, and who may not themselves receive similar benefits in later life because the welfare system looks increasingly unaffordable.[11]

Right-wing advocates see UBI as a way to remove swathes of government bureaucracy: abolishing means testing removes the need for the battalions of civil servants who devise and implement it. There would be no incentive for people to game the benefits system, thus reducing government-generated waste and unfairness. They hope it would facilitate a wholesale simplification of tax structures, and perhaps enable a move to a flat tax. And they argue that more lower-income people would go to work because they would no longer be caught in benefit traps which penalise them for raising their income slightly. This would mean fewer children raised in families where nobody works, a particular bugbear of the right.[12]

Most current supporters of UBI are on the left, but it has had support from prominent right-wing politicians and economists in the past, notably President Richard Nixon and economists Friedrich Hayek and Milton Friedman.

14.3.2 Experiments

There have been a surprising number of experiments with UBI: the Basic Income page on Reddit lists 25,[13] and gives potted descriptions of the purpose and outcomes of six of them.[14] All the researchers involved reported excellent results, with the subjects experiencing healthier, happier lives, and not collapsing into lazy lifestyles or squandering the money on alcohol or other drugs. Given that, it is curious that none of the experiments have been extended or made permanent.

The declared purpose of many UBI experiments is to investigate the concern that when people receive money for nothing, they stop working. One of the biggest experiments conducted so far, involving all 10,000 people in the small town of Dauphin in Manitoba, Canada, found that the only two social groups which did stop working were teenagers and young mothers, and this was seen as a positive outcome.[15]

Of course, people handing in their notice will be of no concern if machines have already stolen all our jobs, but a more subtle version of the concern remains: do people in receipt of money for nothing stop doing anything of value? Do they become indolent couch potatoes, watching TV all day long, or collapse into reliance on alcohol and other drugs? Bearing in mind the distinction we made earlier between jobs and work, in a world where intelligent machines have automated most economic activity, the question is not, do people give up jobs, but do they give up work?

Unfortunately, none of the UBI experiments carried out so far constitute a rigorous test. A rigorous test would be universal, randomised, long-term and basic – in the sense that the income distributed should be enough to live on.[16] And so more tests are planned.

In fact, a number of significant UBI experiments are planned or under way at the time of writing. One, in Finland, caused great excitement when it was announced in 2015, but the aims related to the right-wing concerns listed previously. The Finnish researcher in charge of designing it, Olli Kangas, was hoping to demonstrate solutions for three problems with the current Finnish benefit system. First, people working part-time (perhaps in the gig economy) receive neither work-based benefits nor unemployment benefits. Second, some people are caught in a benefits trap whereby as their income increases their benefits decrease, which removes their incentive to work more and contribute more to the economy. Third, the existing benefits system is expensive, requiring too many bureaucrats to administer it.

The sample of Finns who were chosen to receive the UBI are being compared with a control sample who are not. Kangas will be exploring their propensity to continue working, their reported happiness and well-being, and any changes in their use of health and social services. He hoped to recruit a substantial sample – perhaps 100,000, which would enable him to detect variations between people of different ages, locations, demographics and employment histories.[17] In the event, the trial was scaled back to just 2,000 people, who each received a modest €560 per month. No results will be issued until the trial ends in December 2018,[18] but it already clear that the concept of UBI will not be validated or diminished by it.

Another interesting UBI experiment is a crowd-funded initiative in Germany, which was launched by Berlin-based entrepreneur Michael Bohnmeyer in 2014.[19] By December 2015, 26 people had been selected by lottery to receive €1,000 a month, paid for by public donations. Most of the recipients reported that it didn't change their lives enormously, but they felt less stressed, and in many cases were able to embark on creative projects.

There is no shortage of places keen to experiment with UBI. The Dutch cities of Utrecht, Groningen, Wageningen and Tilburg are asking their national government for permission to carry out trials. All these initiatives are looking for ways to tackle problems with existing social welfare systems.

We have to go to Silicon Valley to find an experiment specifically designed to explore the impact of UBI in the context of a jobless future when machine intelligence has automated most of what we currently do for a living. Just such an experiment was announced in January 2016 by Sam Altman, president of the seed capital firm Y Combinator, which gave a start in life to Reddit, Airbnb and Dropbox. The project will select 3,000 individuals at random from two US states. One thousand of them will receive $1,000 per month for five years, and for comparison purposes, the other 2,000 will receive $50 a month.[20] At a total cost of $66 million, this is a serious project, and Altman's task is not trivial: he will have to figure out a way to quantify the satisfaction his guinea pigs derive from their UBI, and whether they are doing anything useful with their time.[21]

14.3.3 Socialism?

With all these experiments bubbling up, the concept of UBI has become a favourite media topic, but it is controversial. Many opponents – especially in the United States – see it as a form of socialism, and the United States has traditionally harboured a visceral dislike of socialism. (The strong performance of Senator Bernie Sanders, a self-proclaimed democratic socialist, in the race to become the Democratic Party's candidate in the 2016 presidential election is a striking departure from this norm of US politics.)

Martin Ford's book, *The Rise of the Robots*, expressed the hope that UBI was part of the solution to technological unemployment, but the hope was tarnished by his fear that 'guaranteed income is likely to be disparaged as "socialism"', and introducing it will be a 'staggering challenge'. Ford is not alone: I have heard similar concerns from a number of thoughtful American friends.

These fears may be exaggerated. America is, of course, huge – more a continent than a country – so generalisations about it are dangerous. But its people are not in general un-thinking or malicious. If and when it becomes impossible to deny that a majority (or even a large minority) of its citizens will never do paid work again, and for no fault of their own, it is unlikely that the rest will simply allow them to starve.

The dramatic recent changes in American attitudes towards homosexuality and drugs show how fast opinions there can change, and how far. As recently as 1962, homosexual acts were illegal in every US state, and it was only in 2003 that the federal Supreme Court decision in the *Lawrence v Texas* case invalidated the ban in the last 14 states where it remained unlawful. (Even today, more than a dozen states have yet to repeal or amend their own legislation to reflect this ruling.[22]) And yet in June 2015, the federal Supreme Court ruled that bans on same-sex marriage are unconstitutional, in the case of *Obergefell v Hodges*. According to a *Wall Street Journal* poll, public support for gay marriage has doubled in the last decade, standing now at 60%.[23]

Attitudes towards the legalisation of cannabis have also undergone a rapid sea change. For years, governments proclaimed a war on drugs, but that policy has clearly failed. Billions of dollars have been spent, and countless lives have been lost, but supply has not been constrained, much less eliminated. Parts of Mexico and other countries where the drugs are grown or routed have become war zones, and hugely powerful criminal organisations have been spawned. Attempts to curb demand have also failed, with tens of thousands of people being criminalised for an activity that harmed no one. Drugs are dangerous, and their supply should be regulated, but ceding control over that supply to criminal gangs has not proved an enlightened policy. Public opinion in America is swinging rapidly towards that position. In 1969, only 16% of voters polled by Gallup supported legalisation, but now a majority takes that view.[24] Possession of cannabis for personal use is now legal in four states, with the federal government agreeing not to interfere.[25]

It is not only America which is experiencing revolutionary changes in social attitudes. Up until 1997, sex before marriage was illegal in China, condemned as 'hooliganism'. Nevertheless, a researcher found in 1989 that 15% of citizens had experienced it. The percentage had risen above 70% by 2014. Homosexuality was illegal until 2001 and gay marriage is still not legal. But in 2011, state-owned media began writing positive articles about gay pride marches in Shanghai and elsewhere.[26]

These examples show that entrenched societal opinions can and do change, sometimes quickly. If and when machine intelligence renders many of us permanently unemployable, it seems reasonable to expect that opposition to some form of payment to the long-term unemployed will evaporate. After all, the alternative would be mass starvation, and almost certainly, social collapse.

14.3.4 Inflationary?

Opponents of UBI also worry that it will stoke inflation. Other things being equal, a massive injection of money into an economy is liable to raise prices, leading to sudden inflation and perhaps even hyper-inflation. But as UBI campaigner Scott Santens points out, UBI does not necessarily mean an injection of fresh cash into the economy. It would most likely be paid for by increased taxation of the better-off, and by replacing the existing benefits system, together with the bureaucracy which implements it.[27] He also claims that where basic incomes have been introduced, as in Alaska in 1982 and Kuwait in 2011, inflation actually fell.

14.3.5 Unaffordable?

An objection to UBI with more substance is that it is unaffordable. Some argue that it can be funded by raising taxes on the small minority who have become extraordinarily wealthy in recent years. After all, even some of those wealthy people themselves (like Bill Gates and Warren Buffet) have confessed to feeling under-taxed.

Very wealthy people do sometimes decide to dedicate much of their wealth to charitable causes. Bill Gates (again) and Mark Zuckerberg are obvious examples, and as we saw in Section 2.2, many of the robber barons of the late nineteenth century gave fabulous sums to charitable foundations.

(The super-rich might decide to surrender much of their wealth in the interest of self-preservation as much as for philanthropic reasons. We might even call the tax the NBFATW tax, the Not Being First Against The Wall come the revolution tax.[28] Alternatively, they might be sceptical about how grateful the rest of us are likely to be, and decide instead to live on floating fortresses, protected by AI-powered defensive systems of awesome capability.)

If people do give their money away, they generally want to determine for themselves how their wealth is deployed, not least because they believe that they will make better use of it than politicians and bureaucrats.

So, even the most generously disposed wealthy people often resist the wholesale appropriation of their assets in the form of taxation. And as demonstrated by the Panama papers scandal that erupted in April 2016, they are well equipped to do so, either by hiring clever lawyers and accountants to find loopholes and dodges, or by shifting themselves and their assets to less demanding jurisdictions.

Furthermore, entrepreneurs and other capable commercial people who are not yet extremely wealthy but aspire to become so may decide to move out of a jurisdiction which raises taxes sharply to pay for UBI. Or if they stay, they may become discouraged and decide against taking the necessary risks and dedicating the necessary time and energy to projects which could achieve their ambition. These people are responsible for much of the dynamism in capitalist countries, and dampening their enthusiasm or incentivising them to move elsewhere can be very damaging to an economy.

This sounds like common sense, but is in fact highly contentious. The political left believes that inequality is a social evil, and argues that taxing the rich does not deter economic activity.[29] The political right believes that a modicum of inequality is no bad thing, and is anyway inevitable in a thriving economy. It argues that increasing taxes on the rich does deter economic activity, and may actually result in lower government revenues, as the rich look harder for ways to reduce their tax burden.[30]

14.3.6 The Laffer Curve

Unfortunately, the data is muddy, which enables both sides to marshal apparently convincing arguments. And as is so often the case, the truth lies somewhere between them. We do know that there is a level of taxation beyond which further increases are ineffective, or even self-defeating. The Laffer curve plots tax rates against the revenue they raise. At 100%, no one would work, so that is an inefficient rate; 99% would not be much better. Sadly, we just don't know for sure what the optimal level is, either in general, or in a specific country at a specific time.[31]

In the United Kingdom, the Labour government in 2010 introduced a top rate of tax of 50% for people earning above £150,000. The Conservative government took it down to 45% in 2013, and claimed the result was a sharp revenue increase. The Labour party, of course, claimed the opposite.[32]

Which side you choose in this debate will be determined largely by your political orientation. Personally, I believe that competing organisations in well-regulated markets are more efficient and effective than monopolistic governments, and I believe that lower tax regimes encourage entrepreneurship. I also think that governments tend to tax their subjects as much as they think they can get away with, which explains why so much of their tax take is achieved through subtle, indirect and often downright stealthy taxes. Thus a substantial tax increase to fund UBI is likely to be economically damaging.

If you are on the political left you are likely to disagree with this. Fortunately, I may not need to tempt you to cross the parliamentary floor (which will be a relief to us both) because the proponents of UBI think they have two potential sources of revenue to pay for the scheme which mean they don't simply have to 'soak the rich' (an expression coined in 1935 when Franklin D Roosevelt raised the top rate of income tax to 75% in order to pay for the New Deal.) These are saving money by eliminating bureaucracy and by taxing robots.

14.3.7 Let's Kill All the Bureaucrats

Channelling Shakespeare,[33] UBI advocates claim that the massive cost of UBI could be offset by abolishing much or all of the existing benefits systems, along with the legions of bureaucrats who implement them. They offer an enticing vision of a world without means testing, with no poverty traps, no steely eyed 'advisers' in job centres forcing claimants to apply for unsuitable work, no benefit fraud and no need to game the system.

Unfortunately, the world probably won't allow such a nice, tidy outcome. People's needs vary according to their capabilities, their life stage and their location, among other things. Someone who is disabled might well suffer greatly if their income was equal to that of an able-bodied person in robust health. A single dad with a child may need extra support. People living in London or San Francisco would certainly need more housing benefit than people living in Albuquerque or Auchtermuchty. Having ushered all the bureaucrats out the door thanks to the purifying simplicity of UBI, we would have to apologise and call them right back in again.

The RSA, a British think tank, published a report about UBI in December 2015 which was the result of a year's research and discussions.[34] It proposed abolishing much of the United Kingdom's existing benefits system, and replacing it with a payment of £3,692 for everyone between 25 and 65. This is £307 a month, £71 a week, or £10 a day. The payment amounts to a modest 14% of the average UK wage, which was £26,500 in 2015.[35] People aged between five and 25 would receive £2,925, and pensioners would get £7,420. Extra payments would be made for young children.

The RSA estimated the total cost of its proposed system at £280 billion, including running costs of £3 billion. It claimed that this would be offset by £272 billion saved by abolishing most of the existing benefits and pensions infrastructure, including personal income tax allowances and tax relief on pensions payments for higher rate tax payers.

The RSA claimed that families with children and on low wages would be £2,000 to £8,000 better off per year because of the removal of benefit traps. Adjustments required to prevent poorer people being worse off would take the cost to between £10 and 16 billion, around 1% of GDP. This would be funded by taxes on high earners, a group which would also lose income from the changes.

The RSA scheme is not a fully-fledged UBI proposal, as payments would taper off for incomes above £75,000, and stop altogether at £100,000. The level of payments are also set at a level which would keep people alive, but would not provide a decent standard of living. It is also significant that the proposal ignores payments for housing and disability, which are of course substantial, and would require the recall of at least some of those bureaucrats.

Like the Finnish experiment, the idea makes more sense as an attempt to simplify and streamline the United Kingdom's messy and Byzantine benefits system.

Countries are not isolated economic ecosystems. Introducing UBI would significantly affect the competitive position of a country which introduced it, and would have other unintended consequences. A surprisingly positive article about UBI in the right-of-centre *Daily Telegraph* newspaper in December 2015 speculated that if Finland's UBI experiment was successful, it will be inundated by economic migrants unless it leaves the EU.[36]

14.3.8 Taxing Robots

If we can't fund UBI by soaking the rich, or by sacking the bureaucrats who operate the existing welfare infrastructure, can we do it by taxing robots?

In an article published in February 2017, Bill Gates floated the idea of taxing robots which replace human workers.[37] The article was accompanied by a short video, at the end of which Gates giggled about the idea of paying more taxes. It got a lot of people very excited, but it is pretty certain that it won't work, and Gates probably understands this. (After all, he is a rather smart fellow.)

Imagine two firms offering the same service. One has been going for a few years, and recently replaced a thousand humans with machines. The other is a start-up, and went straight to machines. The former would be hit by a tax that the latter would escape. Not only is that very unfair, it would simply mean the former would close, and the tax take would disappear anyway.

Machines will rarely replace humans on a one-for-one basis. Humans will disappear from call centres, for instance, and be replaced by an AI system running on large numbers of servers in a big building with great air conditioning. Does the government tax one entity – namely the AI system – or each individual server? Or does it estimate the number of humans who have been laid off by the system and calculate the tax based on that?

In reality, Gates was probably offering a metaphor for how to cope with the economic singularity. If humans are being laid off in their millions by AI systems, somebody somewhere is making a lot of money by providing the technology which makes that possible. If – and it's a very big if – we can stop them skipping off to the Cayman Islands or somewhere else with very low tax rates, then perhaps we can levy taxes on them to pay an income to the people who have been laid off.

Unfortunately, even this more generalised approach has fatal flaws. The technology replaced the humans because it was cheaper. (It may also produce better products and services and do so faster, but lower cost is almost always going to be part of the reason why these systems are adopted.) That means there is less profit being generated overall, which means a smaller tax base. And as the machines get more and more efficient, that tax base erodes faster.

The erosion of the tax base doesn't stop there. If a company makes a big profit by replacing a group of workers (albeit that less money is now circulating through that part of the economy because it has become more efficient), then competitors will attempt to replace it by making better AI systems. Assuming they succeed – which they often will – they will depress the tax base further.

Perhaps Gates saw this clearly too, and perhaps his real purpose was simply to get more people thinking and talking about how to provide the income that people will need after we lose our jobs.

14.3.9 Unaffordable

It's looking hard to fund UBI by soaking the rich, eliminating the bureaucrats or taxing the robots. *Financial Times* journalist Tim Harford wrote that in current circumstances, UBI appeals to three kinds of people: those happy to see the needy receive less income, those happy to see the state balloon (and risk massive capital flight) and those who can't add up.[38]

John Kay, economics professor at the London School of Economics, wrote in March 2017 that 'either the basic income is impossibly low, or the expenditure on it is impossibly high'. To put it more bluntly than Kay does, if UBI was introduced at an adequate level in any one country (or group of countries) today, there would be a giant sucking sound, as many of the richer people in the jurisdiction would leave to avoid the punitive taxes that would pay for it.[39]

14.3.10 One Out of Three Ain't Good

As well as being unaffordable, the concept has another couple of serious problems which feature in its name: universality and being basic.

In 1977, the American singer Meat Loaf released the classic album *Bat out of Hell*, and the phrase 'ear worm' could have been invented for one of the singles from it, 'Two out of Three Ain't Bad'. UBI has three components: it is universal, basic and an income. Two of these three are problematic, and as Meat Loaf might say, one out of three ain't good.

14.3.11 Universal

The first of UBI's three characteristics is its universality. It is paid to all citizens regardless of their economic circumstances. There are several reasons why its proponents want this. Experience shows that many benefits are only taken up by those they are intended for if everyone receives them. Means-tested benefits can have low uptake among their target recipients because they are too complicated to claim, or the beneficiaries feel uncomfortable about claiming them, or simply never find out about them. Child benefits in the United Kingdom are one well-known example. There is also the concern that UBI should not be stigmatised as a sign of failure in any sense.

But in the case of UBI, these considerations are surely outweighed by the massive inefficiency of universality. In a scenario of, say, 40% unemployment, paying UBI to Rupert Murdoch, Bill Gates and the millions of others who are still earning healthy incomes would be a terrible waste of resources.

14.3.12 Basic

The second characteristic of UBI is that it is basic, and this is its biggest problem of all – even worse than its unaffordability. 'Basic' cannot mean anything other than extremely modest, and if we are to have a society in which a very large minority or a majority of people will be unemployable for the remainder of their lives, we have to do better than providing them with a subsistence income.

This isn't just a question of social justice, important though that is. It is also a question of self-preservation for the people who are still earning. A society in the developed world where a large minority of people have gone from a reasonable standard of living to a subsistence income which they know will never get better is a society that is likely to collapse.

The proponents of UBI argue that the payment will prevent us all from starving, and we will supplement our universal basic incomes with activities which we enjoy rather than the wage slave drudgery faced by many people today. But the scenario envisaged here is one in which many or most humans simply cannot get paid for their work, because machines can do it cheaper, better and faster. The humans will still work: they will be painters, athletes, explorers, builders, VR games consultants, and they will derive enormous satisfaction from it. But they won't get paid for it.

14.3.13 Not UBI but Progressive Comfortable Income?

Despite its drawbacks, UBI is at least an attempt to answer the right question, namely how can we all have a great standard of living if and when the machines take our jobs. And the debate about it is positive because it helps draw attention to that question. Perhaps we can salvage the good part of UBI and improve the bad parts. Perhaps what we need instead of UBI is a PCI – a progressive comfortable income. This would be paid to those who need it, rather than wasting resources on those who have no need. And it would provide sufficient income to allow a rich and satisfying life.

We still have to work out how to pay for this income – how to make it affordable. If it can't be paid for by soaking the rich, taxing the robots or eliminating the bureaucrats, maybe it can be done by making all the goods and services we need very, very cheap.

14.3.14 The Star Trek Economy

It is often said that science fiction tells us more about the present than it does about the future. Most science fiction writers are not actually trying to predict the future, although they may go to considerable lengths to try to make the worlds they create seem plausible. Generally, they are just trying to tell an entertaining story, or maybe use the opportunity that the genre offers to explore something about the fundamental nature of our lives. (At its best, science fiction is philosophy in fancy dress.)

But intentionally or otherwise, science fiction does a very important job for all of us when we think about the future: it provides us with metaphors and scenarios. Many of the most popular science fiction stories present dystopian scenarios: think *Terminator, Blade Runner, 1984, Brave New World* and so on. But there are also positive scenarios, and one of the most popular ones is *Star Trek*.

Set in the twenty-fourth century, *Star Trek* presents a world of immense possibility, of interstellar travel, adventure and split infinitives. And a world without money or poverty. In the 1996 movie *Star Trek: First Contact*, Captain Jean-Luc Picard explains that 'Money doesn't exist in the twenty-fourth century. The acquisition of wealth is no longer the driving force in our lives. We work to better ourselves and the rest of humanity'.

This was not a feature of the original TV series, in which there were quite a few mentions of money and systems of credit. But before he died, Gene Roddenberry stipulated that there was to be no money in the Federation.[40]

Although there is no money, the people in the later *Star Trek* stories do compete with each other – for prestige, for approval, for increased responsibility and for career advancement. One of the things that makes James Tiberius Kirk an outstanding starfleet commander is his fiercely competitive nature. He operates in a profoundly meritocratic environment and will sacrifice a great deal to win.

This is not new. Men and women have always competed for pre-eminence within their tribes and societies, and we are continually applying our ingenuity to work out new ways to do so. Medieval knights risked life and limb for honour and glory, and their descendants fought for national self-determination. Today, many people expend considerable sweat and tears – if less blood – to demonstrate their prowess in writing elegant open-source software, or edit Wikipedia pages.

Money is not required in *Star Trek*'s United Federation of Planets because energy has become essentially free, and products can be manufactured in so-called replicators, devices which create useful (including edible) objects out of whatever matter is available.

Another popular science fiction series with a broadly optimistic (if darkly humorous) outlook is the late Iain M. Banks' *Culture* books, set in a distant future when a technologically advanced humanity has colonised swathes of the galaxy, and enjoys mostly peaceful relations with a host of alien civilisations. The humans are kept company and aided by vastly superior and extraordinarily indulgent machine intelligences, and they lead lives of perpetual indulgence. As Banks put it in a 2012 interview, 'It is my vision of what you do when you are in a post-scarcity society, you can completely indulge yourself. The Culture has no unemployment problem, no one has to work, so all work is a form of play'.[41]

14.3.15 Abundance

The *Star Trek* economy is the post-scarcity economy, the economy of radical abundance. In their 2012 book *Abundance: The Future Is Better than You Think*, Peter Diamandis and Stephen Kotler argue that this world is within reach in the not-too-distant future, thanks largely to the exponential improvement in technology.

Could they be right? Could the solution to the income problem raised by the economic singularity be that all the goods and services that we need for a rich and fulfilling life (and emphatically not a basic one) are produced so efficiently by machines that they become almost free?

At first blush, it sounds implausible, but with a little imagination you can see how it might be possible. Start with the fact that much of what we spend our money on these days is digital, so it can be duplicated and distributed for free. Most of us spend time, effort and money identifying and procuring the media we want: news, information, films, TV, music, books. Much of this is already free, and some of it costs only a small subscription: for a modest monthly fee, Netflix gives you a treasure trove of movies and TV, and Spotify gives you most of the world's music. Much of the cost is the payments to the creators of the original content, and if no one had to earn a living, that cost could reduce substantially.

The next step is to take seriously the idea that energy should become a great deal cheaper this century. At the moment, despite periodic claims by the green lobby that solar power is competitive with fossil fuels, we tax the latter and subsidise the former. But solar power generation is on an exponential downwards cost curve, and our technology for storing and transporting it is improving in leaps and bounds. Replacing fossil fuels with electricity generated by sun and wind isn't going to happen as soon as we all should want, but it will happen.

Once you remove the cost of drivers and fuel from vehicles, transportation can become very cheap. Switching from internal combustion engines to electric power reduces cost considerably: there are fewer moving parts and no wear and tear from combustion. If the future of transport is fleets of self-driving electric cars summoned by apps and owned by their manufacturers rather than their passengers, the cost of getting from A to B could become trivial. At times, Uber has managed to make the cost of its service cheaper than public transport in some big cities, and this is while using human drivers.[42]

Houses and other buildings, furniture, clothing and food seem to be less susceptible to this sort of cost elimination. Eventually, nanotechnology may give us something like the replicators in *Star Trek*, and goods really can become free. That seems like a distant dream today, but long before then, with very cheap energy, very efficient machines and the removal of expensive humans from the production process, we might be surprised by how inexpensive it could become to lead a fulfilling life.

Agriculture provides a nice example of how much scope there is for AI and related technologies to improve the efficiency of our production methods. For reasons of cost as well as good land husbandry, farmers are keen to minimise the amount of pesticides and herbicides they use to keep insects and weeds at bay. Agricultural equipment manufacturer John Deere paid $350 million for Blue River Technology, a company which uses computer vision to identify whether an individual plant is a weed or part of the crop. Weeds are squirted with precision blasts of pesticide, and the expensive chemical is applied only where it is needed.[43]

If this post-scarcity economy of abundance is actually possible, can we transition to it without major social upheaval, and possibly collapse? Peter Diamandis talks of building a bridge to abundance, and this may turn out to be the most important challenge for humanity in the first half of this century.[44]

It is conceivable that our economy would evolve towards abundance if left to its own devices. The whole point of automation is to reduce cost while maintaining or improving quality. But we cannot guarantee it and we should not rely on it.

14.3.16 Staying Innovative

Technological unemployment does not mean that everybody is jobless. It is likely that for a long time – probably at least until the time when (and if) superintelligence arrives – there will be jobs for some humans, and perhaps for many of us. It is likely that those of us still in work will still be operating in some form of competitive market economy. This book is not arguing for fully automated luxury communism – the automation of everything and common ownership of that which is automated.[45]

The genius of the market economy – the principal reason why it is so effective – is that decisions are taken by the people best qualified to take them. The market enables (indeed obliges) each of us to provide truthful signals about what we do and don't want, what we do and don't value. We buy this car and not that car because we prefer it (given our budgetary constraints) and there is no doubt that we are providing a correct signal because we are spending our own money.

Suppliers are highly incentivised by competition to respond to these signals by offering the best possible goods and services. Although it involves a certain amount of waste, as rivals duplicate facilities to develop and produce their output, the competition between them spurs innovation, and raises quality more effectively than any other system we have tried so far.

By contrast, when decisions are made in a centralised, planned economy, somebody is guessing about what is wanted and needed at every level below them. However good their data collection system, and however well-intentioned they are, they will always be out of date and they will often be just plain wrong. There is also a very good chance that corruption will set in, because it is so easy for that to happen. With apologies to Lord Acton,[46] power corrupts, absolute power corrupts absolutely, and corruption is absolutely central to centralised planning.

The economic singularity will not take place overnight. It will probably be a process spanning decades. In the early periods, when large numbers are unemployed but many are still working, we should retain the innovation-sponsoring, wealth-creating mechanism that free markets and capitalism provide. But it will be a different form of capitalism, with a much larger welfare component.

One vision of how it might all work draws on a suggestion that has been around for decades: micro-payments for the provision of data. Data is essential to most machine learning systems, and the tech giants have built their lucrative empires partly by capturing lakes of it, and in many cases keeping it very much to themselves. Arguments that this is a rip-off are hard to sustain, since we value their services so highly that very few of us protest. Those who do protest are free not to use the services, but the vast majority of us do not choose that option.

At the moment, any payments we might receive for our data would probably be nugatory, and would not contribute meaningfully to our lifestyles. But if the prices of most goods and services fall dramatically, that equation might shift. The claim that data is the 'new oil' seems overdone today, but it might become more relevant in the society of radical abundance.

14.3.17 A Footnote about Assets

This section has focused on income as opposed to wealth – for a reason. Most people have little wealth, and are therefore dependent on income. A poll published in January 2015 by a US personal finance website[47] echoed the finding a year earlier by the Federal Reserve[48] that two-thirds of Americans had savings equal to less than three months income. Half of them could not cover an emergency expense of $400 without going into debt. This was aggravated by the recession which began in 2007: the average American family's net worth fell from $136,000 in 2007 to $81,000 in 2013.

Income is the important metric when considering the economic singularity, not wealth. Wealth inequality is far more extreme in today's world than income inequality, both globally and within individual nations. It is also much less significant.

The charity Oxfam created a stir in January 2014 by claiming that the richest 85 people own as much as the poorest 50% of the world.[49] It was repeated so widely and was so helpful in fundraising that Oxfam has repeated it every year since then. The figure may or may not be correct, but it is highly misleading, and it is disingenuous of Oxfam to keep trotting it out each year. Lack of wealth is not the same as poverty. A young professional in New York living a life of luxury and excess may have no net assets, but it would be perverse to describe her as poorer than a North Korean peasant with no debt and a net worth of a few plastic utensils. Furthermore, if it was possible to eradicate this wealth inequality it would not address poverty. If the richest billionaires gave their wealth to the poorest half of the world, it would amount to a one-off payment of around $500 each.[50]

Nevertheless, if you are one of the lucky minority with substantial net assets, you might be wondering how you will be affected if and when technological unemployment takes hold. Will your house be worth more or less in the new economy? How about your vintage Aston Martin or your collection of fine wines? Until and unless we move to a completely different kind of economy, it is likely that some of the wealthy people – especially those who control the AI which creates most of the added value – will remain wealthy, and perhaps become even more wealthy. Perhaps the prices for Stradivarius violins and prime real estate will continue to rise – for sometime at least.

What about the holdings of the much larger number of people in the middle – people who have net assets of a few tens or hundreds of thousands of dollars, perhaps up to as much as a million or two? Unless we switch quickly and smoothly to a very generous form of welfare, it seems likely that the price of assets typically owned by these middle-class people, such as suburban houses and mass-produced cars, will slide as their owners try to replace lost income by liquidating their property. This could happen quickly, as people look ahead, see what is coming, and decide to cash in before the slide starts in earnest. Asset prices are notoriously hard to predict because they depend on events which cannot be foreseen, and also upon perceptions about what may happen, and perceptions about those perceptions. This is yet another good reason why we should be thinking seriously about these matters sooner rather than later.[51]

14.4 ALLOCATION

14.4.1 The House on the Beach

Imagine we manage to cross Peter Diamandis' bridge to an economy of abundance. We are all living comfortable and fulfilling lives. Most of us take advantage of the almost-free goods and services which are provided by the machines of loving grace, and any extra costs we have to meet are funded by (non-punitive) taxes on the income and assets of those who are still working. Those still working are presumably earning good incomes – perhaps very good incomes.

Not everything can be free or nearly free, and many items will always be scarce. There is a finite and regrettably small supply of large houses on empty white sand beaches fringed with palm trees leading down to a turquoise sea. Or penthouse apartments on Manhattan's Fifth Avenue. There is a very small supply of Vermeers and Aston Martin DB5s.

How will we allocate those goods and services which cannot be rendered free or nearly free? At first sight, the answer seems simple and obvious: we will still have money, and we will still have the market. Supply and demand will continue to operate like before.

But most people will have very little money, and hence very little opportunity to acquire the rare and expensive items. And well below the level of Fifth Avenue apartments, the quality of the assets that we use and enjoy will be extremely varied. In every village, town and city, there is a wide variety of quality of housing stock.

The society of abundance will generate egalitarian incomes for the large numbers of people who are not employed, but a decidedly un-egalitarian asset base. Everyone's access to new goods and services will be more-or-less the same, at least within a given territory or jurisdiction, but some people will be living in nice big houses in the posh part of town, while others will be living in small flats with no sound-proofing in grubby apartment blocks in the unfashionable zone.

Will it be like a game of musical chairs? Prior to the economic singularity we all work hard to improve our lot, and then when the machines take our jobs, the music stops and we all sit down in the chairs we have arrived at? And just stay there forever? That seems neither fair nor sustainable.

Do we decide that no one can own the precious things? Perhaps we could turn all the nice houses into museums and keep the scarce movable objects on display there, to be visited (and perhaps used) on payment of a fee, or by scheduled appointment.

In the meantime, we set the efficient machines to revitalising and/or replacing the stock of lower-quality houses, (and cars, and boats, and furniture, and clothes etc.) But it will take a very long time indeed to build a nice new house for everyone who doesn't start off with one. And even when we have completed that gargantuan task, some houses will still be in much nicer places than others.

And who will decide what the cut-off point is between a house which people can carry on living in, and one which is too nice to be private property?

This is the allocation problem.

14.4.2 VR to the Rescue?

In February 2016, when Palmer Luckey and John Carmack were the key executives of Oculus VR, a Facebook-owned manufacturer of VR hardware and software, they talked about a 'moral imperative' to make VR available to us all.[52]

'Everyone wants to have a happy life, but it's going to be impossible to give everyone everything they want.... Virtual reality can make it so anyone, anywhere can have these experiences'. 'You could imagine almost everyone in the world owning [good VR equipment] … This means that some fraction of the desirable experiences of the wealthy can be synthesized and replicated for a much broader range of people'.

Other people have thought about these questions, and not everyone is delighted by the suggestion that VR can assuage the frustration caused by scarcity. Some people think it impossible, and others think it possible but degrading.

The Harvard political philosopher Robert Nozick described a thought experiment back in 1974 featuring an 'experience machine' which could recreate any sensation you choose. Your brain is persuaded that the experience is real, which means that you believe it too, but in fact your body is lying in in a flotation tank, deprived of all sensory input while your brain is hooked up to the machine. Philosophers do a lot of their work by investigating their intuitions, and Nozick's intuition was that no one would use this machine because we value reality too highly. I find it surprising that he came to that conclusion back in 1974, and it would be an even more surprising conclusion to reach today, when so many people spend so much of their lives in simulated realities, albeit only imperfectly simulated. Certainly, a great deal of money is being invested by smart people in the belief that we will consume VR avidly. Nozick died in 2002, so he won't have to find out for himself – maybe he would be relieved.

Other critics see the Oculus founders' view of the future as possible but frightening. Ethan Zuckerman is director of the Massachusetts Institute of Technology (MIT) Centre for Civic Media, and thinks that 'the idea that we can make gross economic inequalities less relevant by giving [poor people] virtual bread and circuses is diabolical and delusional'. Jaron Lanier is a computer scientist and writer who founded VR pioneer VPL Research, and is generally credited with popularising the term 'virtual reality'. He lambasts as 'evil' the vision that the rich will become immortal, while 'everyone else will get a simulated reality … I'd prefer to see a world where everyone is a first-class citizen and we don't have people living in the Matrix'.

Only time will tell if VR is helpful, or even necessary, in enabling us to live in a world where machines have made humans unemployable. My own guess is that it will play a major role in the lives of most people, and that it will make them more productive, more fun and more fulfilling. As Oculus' John Carmack puts it, 'if people are having a virtually happy life, they are having a happy life. Period'.

Nevertheless, Zuckerman and Lanier have identified an important problem with the vision. It is not to do with VR so much as the potential separation of our species into two or more divergent camps. We will review this in more detail in the next section.

14.5 COHESION

14.5.1 The Scenario of 'the Gods and the Useless'

In his remarkable book *Homo Deus*, Yuval Harari makes the brutal suggestion that sooner or later, most people will be unemployable, and the consequence will be a two-tier economy consisting of the gods and the useless. 'As algorithms push humans out of the job market, wealth might become concentrated in the hands of the tiny elite that owns the all-powerful algorithms, creating unprecedented social inequality. ... The most important question in twenty-first-century economics may well be what to do with all the superfluous people'.

Imagine a society where the great majority of people lead lives of leisure, their income provided by a beneficent state, or perhaps a gigantic charitable organisation. They are not rich, they don't travel first class or frequent expensive restaurants, and they don't own multiple houses. But they have no pressing needs and in fact they want for little: they enjoy socialising, learning, sports, exploration, and much of this is carried out in virtual worlds which are almost indistinguishable from reality.

A small minority of people in this society do have jobs. Their work is pleasurable and intellectually stimulating, and not stressful. It involves monitoring and occasionally guiding or resetting the performance of the machines which run their society – machines which they own.

Let's say that this elite minority is generous towards the majority which lives outside their gated communities, and which does not visit the luxurious resorts they migrate between, and does not travel with them on their private heli-jets. They are effectively benign rulers, although both camps refrain from putting it like that.

Now, in this future world, all members of the species of *Homo sapiens* are changing. They are using new technologies to enhance themselves both cognitively and physically. They use smart drugs, exoskeletons and genetic technologies, among others. Maybe they have engineered themselves to need less sleep.[53]

Everyone has access to these technologies, but the elite has privileged access. They get them sooner, and this could be vitally important. I argued in Chapter 4: Tomorrow's AI, that concerns about a 'digital divide' were exaggerated. Companies make much more money by selling lots of relatively cheap cars and smartphones to almost everyone than they ever could by selling just a handful of diamond-encrusted versions to the super-rich.

But we mustn't forget that technology is advancing at an accelerating rate. In the future society we are envisioning, important breakthroughs in physical and cognitive enhancement are announced every year, then every month, then every week. As AI gets better and better it fuels this improvement – even though it is still narrow AI, and far from becoming human-level, artificial general intelligence or AGI.

It may become hard or even impossible to disseminate these cognitive and physical improvements quickly enough to avoid a profound separation between those with privileged access to them and the rest of us.

So, the elite will change faster than the rest. As the two groups lead largely separate lives, the widening gap may not be apparent to the majority, but the elite surely will know about it. They will decide that they must draw attention away from the fact, and they will take precautions to prevent attack, in case the majority should become aware of and resentful about what is happening. They will surround their gated communities with discreet machines which possess astonishingly powerful defensive – and offensive – capabilities. They will keep themselves more and more to themselves, meeting members of the majority less and less often. When they do meet, it will almost always be in VR, where their avatars (their representations in the VR environment) do not betray the widening gulf between the two types of humans.

When normal people read about the lives of billionaires and movie stars, we often think they live in a different world. But the distance which separates them from us is tiny compared with the gulf which could open up between the AI-owning elite and the unemployed majority in a world which passed through the economic singularity while retaining private property.

After a while, humans may speciate: they may evolve into two different species: the gods and the useless.

14.5.2 Brave New World

This is not your average science fiction dystopia. The scenario most commonly offered up by Hollywood is that technology has pulled back the curtain which concealed the truly evil nature of the capitalism system, and mankind has fallen into a form of high-tech slavery, where the rich descendants of company CEOs and scheming tycoons and financiers brutalise an impoverished and oppressed majority.

The movie *Elysium* is just one of many dreary examples of this tired old cliché. In it, as in so many others, society has actually regressed from capitalism into a sort of techno-feudalism. Any viewer who is half-awake is wondering, since machines can do all the work, what is the point of enslaving humans?

(It is curious that the left-leaning culture prevalent in Hollywood impels it to issue these tirades against capitalism, when Hollywood studios are themselves formidable exponents of the capitalist art. And it is curious that Hollywood stars who demand millions of dollars to make a single film complain that corporations are fuelled by nothing but greed.)

A far more interesting scenario is Aldous Huxley's *Brave New World*, which is a very subtle piece of world-building. Almost everyone is content in the society which has been developed after an appalling military conflict, but it is clear to the reader that humanity has lost something important, and most of it has regressed to an almost infantile condition. Yet when a talented outsider arrives, he is unable to devise a way to improve the system, or to accommodate himself to it. He instinctively feels – and the reader is encouraged to agree – that his own life has shown him there is a better way to live, but he is unable to articulate or maintain it.

In Huxley's story, humanity has achieved a stable equilibrium. Only a tiny minority of humans are aware of what has been sacrificed in order to achieve a society of such docility and acquiescence. Regular sex and a powerful drug called soma are the opiates of the masses. (Huxley wrote the book in 1931, before the advent of rock and roll, so he couldn't include the third leg of Ian Dury's hedonistic mantra that sex and drugs and rock and roll were all his brain and body needed.[54])

Brave New World is certainly not intended as a blueprint. Even so, it is calm, and the spectre of economic and social collapse seems to have been abolished – at least for the time being. It would be foolish of us to take even that much for granted. A society comprising gods and the useless might turn out to be inherently unstable. In a full-on conflict between them it seems likely that the gods would have the means to protect themselves, but at what cost?

14.5.3 Will Capitalism Remain Fit for Purpose?

Private property is an essential feature of capitalism, and in particular, the private ownership of the means of production, exchange and distribution. In market economies, most people earn their living by selling their labour – their time and their physical and intellectual skills.

People called entrepreneurs hire workers and combine their labour with the other major element of the capitalist economy – capital, which consists of money, machinery, land, buildings and intellectual property. The entrepreneurs use the labour and the capital to develop and make products and services which they hope will be bought by enough people to turn a profit. The capital is put at risk during this process, and the profit is a reward and an incentive for the entrepreneurs and the owners of the capital (known as capitalists) who took that risk.

The picture is complicated, because we are all capitalists today. Much of the capital deployed by entrepreneurs is owned by financial institutions like insurers and pension companies, and ownership of these institutions is very widely distributed through pension plans. Many people are also shareholders in the companies they work for, thanks to employee share ownership schemes.

In a capitalist economy, where most people work, individuals who start off without any capital can acquire it by saving some of what they are paid for their labour, and by starting companies themselves. Countries where this is easy and celebrated tend to be better off than countries where entrepreneurship is discouraged, or hampered by corruption, over-regulation or lack of infrastructure.

Similarly, people who start off with capital can lose it, by bad luck or poor judgement. It is not uncommon for families to go from rags to riches and back to rags within three generations.

There are countries today where this kind of social mobility (both upward and downward) is very limited, and the elites are entrenched.

Arguably, nowhere in the world has sufficient economic social mobility, but in many countries it does exist to some degree.

As we have seen, after the economic singularity there will be two major differences. First, if the prediction that most people will be unemployable comes true, then there will be pretty much no traffic – no economic and social mobility – between the elite and the rest. This will be the case all over the world. If you can't do paid work, it is very hard to accumulate capital. Second, if the rate of technological progress continues to accelerate, the elite may avail themselves of the means of cognitive and physical enhancement to diverge from the majority, both physically and cognitively.

The obvious but difficult remedy for this is to end the institution of private property. The means of production, exchange and distribution would be placed into some kind of collective ownership to prevent the possibility of social and species fracture.

This conclusion makes me extremely uncomfortable. I was in business for 30 years before becoming a full-time writer and speaker, and I remain convinced of the largely positive effects of a regulated market economy with a welfare safety net.[55] It seems to me that capitalism is the best economic system we know of for a society where humans do the work.

But capitalism may not be so appropriate for an economy of abundance, where machines do the work, where most people are unemployed, and where technology is changing the species quickly. I am not pushing this argument hard. Not only does it make me uncomfortable, but one of the fundamental characteristics of a singularity is that it is even harder than usual to predict the future when there is an event horizon in the way.

It is by no means certain that abandoning capitalism and private property is the only way to avoid fracture and collapse. It may be possible for all kinds of people to live in harmony in a society where a minority gets paid to work and owns most of the economy's assets, while everyone else lives happily on a universal basic income. It is not impossible to imagine a benign technocracy in which the majority of people really don't care who owns what, because they are wholly satisfied with their abundant supply of material and digital goods and services.

Unfortunately, every time I try to envisage this world, the picture degrades into a variation on the theme of *Brave New World* – or worse. Perhaps this is simply a failure of my imagination. I hope so.

If it is true that we need to move away from capitalism, we have two major jobs on our hands. First, we need to determine what that new economy should look like. Second, we need to work out how to transition from the economy we have to the economy we need. This will not be easy. Humans dislike change, and as always, there will be winners and losers. The losers may not take their losses calmly.

14.5.4 Collective Ownership

So perhaps navigating the economic singularity successfully requires the AI-owning elites to transfer their assets into collective ownership, and be hailed as heroes and heroines for doing so. How might this work in practice?

I argued previously that planned economies have inherent flaws, and that part of the genius of the market economy is that decisions are taken by the people best qualified to take them. Common ownership can work well in small communities, such as families, tribes and small villages. But as soon as a society attains any level of size and sophistication, the bonds of kinship weaken and individuals start to claim ownership over land and property. The society becomes regulated by power structures which begin as means of self-defence and evolve into expressions of ambition.

If (and it is a big 'if') surviving the economic singularity and avoiding fracture means ending the system of private ownership, how can this be done without falling into the unwelcome embrace of an over-mighty state and centralised planning?

The answer just might be the blockchain.

14.5.5 Blockchain

Initial coin offerings (ICOs) are a way to raise money for digital currencies and other enterprises employing blockchain technology. In June 2017, the amount of money raised in ICOs overtook the amount invested by the venture capital industry.[56] With the great majority of people still very hazy about blockchain technology, and regulators increasingly nervous that ICOs enable Ponzi schemes and money laundering on a huge scale, this was a remarkable milestone.

The biggest and most famous application of blockchain technology is bitcoin, which launched as open-source software in 2009, having been described in a paper published in 2008 under the pseudonym Satoshi Nakamoto. Its valuation has recorded a rapid and very volatile rise. In March 2017, it reached $1268, overtaking the value of an ounce of gold for the first time. In August 2017, it passed $4500.[57]

The blockchain is a public ledger which records transactions. The important thing is that the ledger is completely trustworthy despite having no central authority, like a bank, to validate it. It is trustworthy in that you can have full confidence that if someone gives you a bitcoin, then you do own that bitcoin: the person who gave it to you will not be nipping off to spend the same piece of currency elsewhere, even though it is entirely digital.

This confidence arises because transactions are recorded in blocks which are added to the chain by people (or rather computer algorithms) called miners. These miners are working continuously on mathematical problems whose solutions are hard to find but easy to verify. A problem is solved ('mined') every few minutes, and each solution creates a block. The new block is added to the chain, and incorporates the transactions made since the last block was added to the chain. Your transaction is published on the blockchain's network as soon as it is agreed, but it is only confirmed, and hence reliable, when a miner has incorporated it into a block.

Satoshi Nakamoto's innovation solved a previously intractable challenge in computer science known as the Byzantine generals' problem. Imagine a medieval city surrounded by a dozen armies, each led by a powerful general. If the armies mount a coordinated attack, their victory is assured, but they can only communicate by messengers on horseback who visit the generals one by one, and some of the generals are untrustworthy. The blockchain provides a way for each general to know that a message calling for an attack at a particular time is genuine, and has not been fabricated by a dishonest general before it reached him.[58]

Digital currency is only one of the possible applications of blockchain technology. It can register and validate all sorts of transactions and relationships. For instance, it could be used to manage the sale, lease or hire of a car. When you take possession of a car, it could be tagged with a cryptographic signature, which would mean that you are the only person who could open and start the car.[59]

The revolutionary benefit of the blockchain is that all kinds of agreements can be validated without setting up a centralised institution to do so. By removing the need for a central intermediary, the blockchain can reduce transaction costs, and it can enhance privacy: no government agents need have access to your data without your permission.

Most importantly, for our present purposes, the blockchain may make possible the decentralised ownership and management of collective assets.

14.5.6 The Elite's Dilemma

Imagine a future in which it is apparent to many people that we are heading towards the scenario of the gods and the useless. The elite few who own the machines are as uncomfortable as the rest of us about this – or at least a sizeable number of them are. They do not want to hand their assets over to a government organisation, as they believe this would simply swap one potentially dangerous elite for another one.

But they realise that if the scenarios of the gods and the useless becomes reality, they will end up as pariahs, feared and perhaps hated by the rest of their species. This outcome might well be grim for the useless, but it would be unpleasant for the gods as well.

I don't believe the common meme that rich people are all bad, greedy and selfish. I have known quite a few, and worked for some of them. They seem to me to be the same mix of good and bad, greedy and generous as the rest of us. They tend to be smart and hard-working, but otherwise they are pretty normal, which is to say, that curious human blend of similar and different, happy and sad, predictable and unpredictable.

It strikes me as entirely plausible that in the gods and the useless scenario, the smallish minority which owns most of the assets when the game of musical chairs stopped – including notably the AI – would prefer to throw in their lot with the rest of us rather than hide behind heavily fortified gates, outcasts from the rest of their species.

It would be a non-trivial project to work out in detail how the assets could be transferred into universal common ownership, validated by the blockchain, and managed in a decentralised fashion. And it is certainly not a forgone conclusion that the rich minority would endorse it. But I suspect it will turn out to be our best way forward.

Of course we will not need to overcome this challenge unless we can overcome the income challenge first. And the income challenge will in turn probably be presaged by a major hurdle: panic.

14.6 PANIC

14.6.1 Fear

Franklin D Roosevelt was inaugurated as US president in March 1933, in the depth of the Great Depression. His famous comment that 'We have nothing to fear but fear itself' was reassuring to his troubled countrymen, and has resonated down the years. In the economic singularity, fear will not be our only problem, but it may well turn out to be our first very serious problem.

Fully autonomous, self-driving vehicles will start to be sold during the coming decade – perhaps in five years, perhaps in ten. Because of the substantial cost saving to the operators of commercial fleets, the humans driving taxis, lorries, vans and buses will be laid off quickly during the decade which follows, so that within 15 or 20 years from now, it is likely there will be very few professional drivers left.

Well before this process is complete, though, people will understand that it is happening, and that it is inevitable. Most of us will have a friend, acquaintance or family member who used to be a professional driver. And the technology that destroyed their job will be very evident. One of the interesting and important things about self-driving cars is they are not invisible, like Google Translate, or Facebook's facial recognition AI systems. They are tangible, physical things which cannot be ignored.

Most people are not thinking about the possibility of technological unemployment today. They see reports about it in the media, and they hear some people saying it is coming, and others saying it cannot happen. They shrug – perhaps shudder – and get on with their lives. This response will no longer be possible when robots are driving around freely, and human drivers are losing their jobs. This cannot fail to strike people as remarkable. Learning to drive is a difficult process, a rite of passage which humans are only allowed to undertake on public roads when they are virtually adult. The fact that robots can suddenly do it better than humans is not something you can ignore.

No doubt some people will try to dismiss the phenomenon by explaining that driving wasn't evidence of intelligence after all: like chess, it is mere computation. Tesler's theorem – the definition of AI as whatever we cannot yet do – will cling on. But most people will not be fooled. Self-driving vehicles will probably be the canary in the coal mine, making it impossible to ignore the impact of cognitive automation. People will realise that machines have indeed become highly competent, and they will realise that their own job may also be vulnerable.

If we have a Franklin D Roosevelt in charge at the time – perhaps one in every country – this may not be a problem. If there is a plausible plan for how to navigate the economic singularity, and a safe pair of hands to implement the plan, then we may be OK. Unfortunately, we do not currently have a plan. There is no consensus about what kind of economy could cope with a majority of the population being permanently unemployed, nor how to get from here to there. Neither are all the top jobs in safe pairs of hands.

In the absence of a solid plan explained by a reassuringly competent leadership, the reaction of large numbers of people realising that their livelihoods are in Jeopardy is not hard to predict: there will be panic.

When will this panic occur? Within a few years, and perhaps within a few months, of self-driving vehicles starting to lay off human drivers. In other words, in a decade or so.

14.6.2 Populism

Populist is a title applied to politicians who claim to represent the common people in a perceived battle against the interests of an established elite. Few people actively claim the title, as it also denotes a politician who promises much but will deliver little, and will probably cause great harm to those who elected him or her, as well as to everyone else.

In the first sense at least, most observers would agree that the election of President Trump and the British referendum vote to leave the EU were the result of populist campaigns. (Let's not get into a row about whether those political movements also deserve the second part of the title.)

It is usually thought that the success of these campaigns was due to a feeling among voters that they have been 'left behind' and economically disadvantaged since the Great Recession of 2008 onwards. In which case, it is odd that populists had their greatest successes in countries which performed better economically than most others. US GDP was 0.26% higher in 2016 than in 2008, which is hardly a sparkling performance, but it is considerably better than the EU, whose GDP shrank by 0.14%.[60] The United Kingdom's GDP grew by 0.8%, just a whisker behind the most successful large European economy, Germany, whose GDP grew by 0.9%.

Why did the relatively successful United States and United Kingdom experience extreme populism, when the less successful France (down 0.16%), Italy (down 0.23%) and Spain (down 0.25%) did not?

There is a clue in the most prominent slogan of each campaign: 'Make America Great Again', and 'Take Back Control'. Trump and Brexit were backward-looking, nostalgic campaigns which promised a return to a better time that supposedly existed in the past. Yes, they were campaigns against allegedly corrupt and smug elites (Wall Street and Washington in the United States, 'the metropolitan elite' in the United Kingdom, and the national broadcast media in both), but first and foremost they were campaigns against change.

What change were they seeking to reverse? As the West's most open economies, and among their most successful, the United States and the United Kingdom attracted more than their fair share of young, ambitious expatriates since the 2008 crash. In the early 2010s, if you were young, entrepreneurial and Polish, Greek, Mexican or Vietnamese, you would be quite likely to decide that your future was more promising in London, New York or San Francisco than in your home country. Once there, if not before, you would probably adopt a left-leaning, politically correct world view, and this, together with the simple fact of being foreign, would make you an object of suspicion in the eyes of a native population that was less open to change, and feared your impact on its employment prospects and access to public services.

14.6.3 Panic

The election of Trump and the Brexit referendum result were political earthquakes. Politics has not been so interesting since at least the fall of the Berlin Wall and the end of the Cold War at the end of the 1980s. But compared with a realisation by the majority of the population that they are very likely to lose their jobs, their causes were relatively minor. The possible impacts of a panic about impending widespread joblessness could be enormous, and they are worth expending considerable effort to avoid.

We will explore some of them in the next two chapters, along with some more cheerful scenarios. After that, in Chapter 17: The Economic Singularity, we will look at some recommendations for how to steer ourselves towards the better scenarios.

NOTES

1. In case you only recently arrived on this planet, that was a reference to the sainted Douglas Adam's *Hitchhiker's Guide to the Galaxy* series. If you haven't read it, I recommend that you put this book down and read that one instead. I won't be offended. But please come back here afterwards.
2. http://philpapers.org/archive/DANHAT.pdf.
3. The novel is sometimes said to have originated in the early eighteenth century, but in fact it is a much older art form. What happened then was that writers began publishing books which described life as they actually saw it. https://en.wikipedia.org/wiki/Novel#18th_century_novel.

4. I am indebted to AGI researcher Randal Koene for this observation.
5. https://en.wikiquote.org/wiki/Bette_Davis.
6. http://www.economist.com/node/17722567.
7. http://quoteinvestigator.com/2011/11/16/robots-buy-cars/.
8. http://thegreatdepressioncauses.com/unemployment/.
9. http://www.statista.com/statistics/268830/unemployment-rate-in-eu-countries/.
10. http://www.statista.com/statistics/266228/youth-unemployment-rate-in-eu-countries/.
11. http://www.scottsantens.com/.
12. http://www.economonitor.com/dolanecon/2014/01/27/a-universal-basic-income-conservative-progressive-and-libertarian-perspectives-part-3-of-a-series/.
13. https://www.reddit.com/r/BasicIncome/wiki/index#wiki_that.27s_all_very_well.2C_but_where.27s_the_evidence.3F.
14. https://www.reddit.com/r/BasicIncome/wiki/studies.
15. http://basicincome.org.uk/2013/08/health-forget-mincome-poverty/.
16. http://fivethirtyeight.com/features/universal-basic-income/?utm_content=buffer71a7e&utm_medium=social&utm_source=plus.google.com&utm_campaign=buffer.
17. http://www.fastcoexist.com/3052595/how-finlands-exciting-basic-income-experiment-will-work-and-what-we-can-learn-from-it.
18. http://basicincome.org/news/2017/05/finland-first-results-basic-income-pilot-not-exactly/.
19. http://www.latimes.com/world/europe/la-fg-germany-basic-income-20151227-story.html.
20. https://www.cnbc.com/2017/09/21/silicon-valley-giant-y-combinator-to-branch-out-basic-income-trial.html.
21. http://www.vox.com/2016/1/28/10860830/y-combinator-basic-income.
22. https://en.wikipedia.org/wiki/Sodomy_laws_in_the_United_States#References.
23. http://blogs.wsj.com/washwire/2015/03/09/support-for-gay-marriage-hits-all-time-high-wsjnbc-news-poll/.
24. http://www.huffingtonpost.com/2009/05/06/majority-of-americans-wan_n_198196.html.
25. http://blogs.seattletimes.com/today/2013/08/washingtons-pot-law-wont-get-federal-challenge/.
26. http://www.bbc.co.uk/news/magazine-35525566.
27. https://medium.com/basic-income/wouldnt-unconditional-basic-income-just-cause-massive-inflation-fe71d69f15e7#.3yezsngej.
28. A splendidly cynical idea from my friend Matt Leach.
29. http://www.forbes.com/sites/greatspeculations/2012/12/05/how-i-know-higher-taxes-would-be-good-for-the-economy/#5b0c080b3ec1.
30. http://taxfoundation.org/article/what-evidence-taxes-and-growth.
31. https://en.wikipedia.org/wiki/Laffer_curve.
32. http://www.bbc.co.uk/news/uk-politics-26875420.

33. A minor character in Shakespeare's *Henry VI* called Dick the Butcher has the memorable line, 'First thing we do, let's kill all the lawyers'. It seems Shakespeare was not fond of lawyers, http://www.spectacle.org/797/finkel.html.

34. https://www.thersa.org/action-and-research/rsa-projects/public-services-and-communities-folder/basic-income/

35. http://www.icalculator.info/news/UK_average_earnings_2014.html.

36. http://www.telegraph.co.uk/finance/economics/12037623/Paying-all-UK-citizens-155-a-week-may-be-an-idea-whose-time-has-come.html.

37. https://qz.com/911968/bill-gates-the-robot-that-takes-your-job-should-pay-taxes/.

38. http://timharford.com/2016/05/could-an-income-for-all-provide-the-ultimate-safety-net/.

39. http://archive.intereconomics.eu/year/2017/2/the-basics-of-basic-income/.

40. http://fee.org/freeman/the-economic-fantasy-of-star-trek/.

41. https://www.wired.co.uk/news/archive/2012-11/16/iain-m-banks-the-hydrogen-sonata-review.

42. http://www.slate.com/blogs/moneybox/2016/07/11/uber_s_manhattan_commute_card_signifies_the_company_s_move_from_taxis_towards.html.

43. https://www.wired.com/story/why-john-deere-just-spent-dollar305-million-on-a-lettuce-farming-robot/?mbid=social_twitter_onsiteshare.

44. https://medium.com/@PeterDiamandis/a-bridge-to-abundance-6d83738d55dd.

45. https://www.theguardian.com/sustainable-business/2015/mar/18/fully-automated-luxury-communism-robots-employment.

46. http://history.hanover.edu/courses/excerpts/165acton.html.

47. http://www.marketwatch.com/story/most-americans-are-one-paycheck-away-from-the-street-2015-01-07.

48. http://www.federalreserve.gov/econresdata/2014-economic-well-being-of-us-households-in-2013-executive-summary.htm.

49. http://www.theguardian.com/business/2016/jan/18/richest-62-billionaires-wealthy-half-world-population-combined.

50. http://www.bbc.co.uk/news/magazine-26613682.

51. I'm indebted to Dr Justin Stewart, an investor, for prodding me to address the issue of assets more closely.

52. http://www.wired.com/2016/02/vr-moral-imperative-or-opiate-of-masses/.

53. http://motherboard.vice.com/read/sleep-tech-will-widen-the-gap-between-the-rich-and-the-poor.

54. https://en.wikipedia.org/wiki/Sex_and_drugs_and_rock_and_roll.

55. I am that terrible old cliché: a socialist student whose left-wing views did not long survive contact with the real world. As a trainee BBC journalist writing about Central and Eastern Europe in the decade before the Berlin Wall fell, I soon realised how fortunate I was to have grown up in the capitalist West.

56. https://www.cnbc.com/2017/08/09/initial-coin-offerings-surpass-early-stage-venture-capital-funding.html.

57. http://www.telegraph.co.uk/technology/2017/08/24/bitcoin-price-stays-4000-will-continue-rise-will-bubble-burst/.

58. http://www.dugcampbell.com/byzantine-generals-problem/.
59. http://www.economistinsights.com/technology-innovation/analysis/money-no-middleman/tab/1.
60. World Bank data: https://data.worldbank.org/indicator/NY.GDP.MKTP.CD?view=chart.

CHAPTER 15

Four Scenarios

15.1 NO CHANGE

In a July 2015 interview with *Edge*, an online magazine, Pulitzer Prize-winning veteran *New York Times* journalist John Markoff lamented the deceleration of technological progress – in fact he claimed that it has come to a halt.[1] He reported that Moore's Law stopped reducing the price of computer components in 2013, and pointed to the disappointing performance of the robots entered into the Defence Advanced Research Projects Agency (DARPA) Robotics Challenge in June 2015 (which we reviewed in Section 2.3).

He claimed that there has been no profound technological innovation since the invention of the smartphone in 2007, and complained that basic science research has essentially died, with no modern equivalent of Xerox's Palo Alto Research Centre (PARC), which was responsible for many of the fundamental features of computers which we take for granted today, like graphical user interfaces (GUIs) and indeed the PC.

Markoff grew up in Silicon Valley and began writing about the Internet in the 1970s. He fears that the spirit of innovation and enterprise has gone out of the place, and bemoans the absence of technologists or entrepreneurs today with the stature of past greats like Doug Engelbart (inventor of the computer mouse and much more), Bill Gates and Steve Jobs. He argues that today's entrepreneurs are mere copycats, trying to peddle the next 'Uber for X'.

He admits that the pace of technological development might pick up again, perhaps thanks to research into meta-materials, whose structure absorbs, bends or enhances electromagnetic waves in exotic ways. He is dismissive of artificial intelligence because it has not yet produced a conscious mind, but he thinks that augmented reality might turn out to be a new platform for innovation, just as the smartphone did a decade ago. But in conclusion he believes that '2045... is going to look more like it looks today than you think'.

It is tempting to think that Markoff was to some extent playing to the gallery, wallowing self-indulgently in sexagenarian nostalgia about the passing of old glories. His critique blithely ignores the arrival of deep learning, social media and much else, and dismisses the basic research that goes on at the tech giants and at universities around the world.

Nevertheless, Markoff does articulate a fairly widespread point of view. Many people believe that the Industrial Revolution had a far greater impact on everyday life than anything produced by the information revolution. Before the arrival of railroads and then cars, most people never travelled outside their town or village, much less to a foreign country. Before the arrival of electricity and central heating, human activity was governed by the sun: even if you were privileged enough to be able to read, it was expensive and tedious to do so by candlelight, and everything slowed down during the cold of the winter months.

But it is facile to ignore the revolutions brought about by the information age. Television and the Internet have shown us how people live all around the world, and thanks to Google and Wikipedia, and so on, we now have something close to omniscience. We have machines which rival us in their ability to read, recognise images and process natural language. And the thing to remember is that the information revolution is very young. What is coming will make the Industrial Revolution, profound as it was, seem pale by comparison.

15.1.1 The Productivity Paradox

Part of the difficulty here is that there is a serious problem with economists' measurement of productivity. The Nobel laureate economist Robert Solow famously remarked in 1987 that 'you can see the computer age everywhere but in the productivity statistics'. Economists complain that productivity has stagnated in recent decades. Another eminent economist, Robert Gordon, argues in his 2016 book *The Rise and Fall of American Growth* that productivity growth was high between 1920 and 1970 and nothing much has happened since then.

Anyone who was alive in the 1970s knows this is nonsense. Back then, cars broke down all the time and were also unsafe. Television was still often black and white, it was broadcast on a tiny number of channels, and it was shut down completely for many hours a day. Foreign travel was rare and very expensive. And we didn't have the omniscience of the Internet. Much of the dramatic improvements that have improved this pretty appalling state of affairs is simply not captured in the productivity or GDP statistics.

Measuring these things has always been a problem. A divorce lawyer deliberately aggravating the animosity between her clients because it will boost her fees is contributing to GDP because she gets paid, but she is only detracting from the sum of human happiness. *The Encyclopedia Britannica* contributed to GDP, but Wikipedia does not. The computer you use today probably costs around the same as the one you used a decade ago, and thus contributes the same to GDP, even though today's version is a marvel compared with the older one. It seems that the improvement in human life is becoming increasingly divorced from the things that economists can measure. It may well be that automation will deepen and accelerate this phenomenon.

The particulars of the future are always unknown, and all predictions are perilous. But the idea that the world will be largely unchanged three decades hence seems the least plausible of the scenarios set out in this chapter.

15.2 FULL EMPLOYMENT

In Chapter 13: Is It Different This Time? We reviewed two arguments against the thesis that cognitive automation will lead to lasting widespread unemployment, and seven examples of ways in which humans can remain employed as robots take our old jobs.

The arguments were what I characterise as the reverse Luddite fallacy and the argument from inexhaustible demand. The reverse Luddite fallacy says that automation has not caused lasting unemployment in the past, and therefore it cannot in the future. When stated as baldly as that, it is clear how weak the argument is: past performance is no guarantee of future outcome, and in any case, automation did cause massive unemployment in the past – of horses.

The argument from inexhaustible demand does not explain how humans can satisfy any of the demand when machines can do most things cheaper, better and faster. If humans and machines chase each other down the revenue curve in satisfying demand, humans will reach the point where it is not worth them continuing long before machines do.

But the fact that the arguments purporting to show that technological unemployment cannot happen are weak does not mean that it will necessarily happen. The fact is that we do not yet know. What I am arguing is that it is certainly possible (indeed likely) and therefore we should prepare for the eventuality since the consequences would otherwise be severe.

The seven examples of how humans will keep working can be summarised as follows:

1. We will form centaur-like partnerships with machines in which they will take care of brute force calculations and lower-level cognitive tasks, and humans will provide the imagination, creativity and flair.

2. The ability to trawl through masses of data and surface all possible correlations will make it possible to carry out tasks which were previously impossible, or were previously unaffordable. We saw this in the case of lawyers becoming able to review thousands of employment records at a much lower price point, and we called it the iceberg effect.

3. Humans will do all the jobs which require empathy, which machines cannot have because they are not conscious.

4. Humans will buy products (and perhaps services) from other humans in preference over goods made by machines because of an

innate chauvinism, or because the human-produced goods have the artisanal quality of being imperfect and slightly different, one from the next.

5. Humans will be entrepreneurs, which machines cannot be because they have no ambition.

6. Humans will be artists, which machines cannot be because art requires the communication of an experience, and unconscious machines can have no experiences.

7. If all else fails, we will open the magic jobs drawer, and out will fly all kinds of new activities that we cannot imagine today because the technology to make them possible has not yet been invented.

As we saw, there are strong counterarguments and counterexamples to each of these, and it seems unlikely they can keep us all in employment. But collectively, they provide comfort to the technological unemployment sceptics, and the self-professed 'deniers', like Google's Eric Schmidt.

15.3 DYSTOPIAS

15.3.1 Fragile – Handle with Care

Civilisation is fragile. Any schoolchild can name some great empires which collapsed: the Greeks, the Persians, the Romans, the Maya, the Inca, the Mughals, the Khmer, the Ottomans, the Hapsburgs. The ancient Egyptians managed to rise and fall several times during their extraordinary 3,000-year history.

We also know how fragile civilisation is from two famous episodes in experimental psychology. In 1961, Yale psychologist Stanley Milgram recruited students from that elite university, and told them to administer mild electric shocks to incentivise a stranger who was supposed to be learning pairs of words. The shocks were fake, but the students did not know this, and an extraordinary two-thirds of the students were prepared, when urged on by the experimenter, to deliver what appeared to be very painful and damaging doses of electricity.[2] The experiment has been replicated numerous times around the world, with similar results.

Ten years later, Stanford psychology professor Philip Zimbardo, a school friend of Milgram's, ran a different experiment in which students were recruited and arbitrarily assigned the roles of prisoners and guards in a make-believe prison. He was shocked to see how enthusiastically sadistic the students who were chosen to be guards became, and he was obliged to terminate the exercise early.[3] This experiment has also been replicated numerous times.

Our twenty-first-century global civilisation seems pretty robust. We have just gone through what is frequently described as the worst recession since the Great Depression of the 1930s, and for the great majority of people, the experience was nothing like as awful as those terrible years, which did so much to set up the disastrous carnage of World War II.

But history and experimental psychology demonstrate that we cannot afford to be complacent. If the argument of Section III of this book is correct, we are about to embark on a journey towards a new type of economy which we have not yet designed. Unless we are careful, there will be plenty of opportunities for missteps, misunderstandings and downright mischief by populists and demagogues.

If technological unemployment arrives in a rush, and we are not prepared, a lot of people will lose their incomes quickly, and governments may not move fast enough to avert drastic collapses in asset prices as people sell their belongings to make ends meet. If the replacement of incomes is slow or botched in some countries, the resulting economic crises could lead to their governments being overthrown by irresponsible or foolish leaders. Among other things, we must hope this does not happen in any countries with significant stocks of nuclear weapons.

In the developed countries, and increasingly, elsewhere too, our lives are intertwined and interdependent. Especially if we live in cities – which more than half of us now do – we depend on just-in-time logistics systems to deliver food and other essentials to our local shops. How long would any of us survive if all the supermarkets suddenly lost their supply chains, and everyone around us was getting hungry?

15.3.2 Preppers

There are signs that the people best-placed to understand what is coming have a keen appreciation that the changes bearing down on us may not end well. Reid Hoffman, co-founder of LinkedIn, told a journalist in early 2017 that half his peer group of Silicon Valley billionaires have some level of 'apocalypse insurance' in the form of a hideaway in the United States or abroad.[4] New Zealand is a favourite location, as what its natives used to call the 'tyranny of distance' becomes a virtue if the rest of the world is going to hell in a handcart.

'Saying you're "buying a house in New Zealand" is kind of a wink, wink, say no more. Once you've done the Masonic handshake, they'll be, like, "Oh, you know, I have a broker who sells old ICBM silos, and they're nuclear-hardened, and they kind of look like they would be interesting to live in"'.

Preppers – people who are preparing for the worst – have their own language. They are preparing for the day when SHTF (the shit hits the fan) and the United States is WROL (without the rule of law). They joke that FEMA, the Federal Emergency Management Agency, actually stands for foolishly expecting meaningful aid.

It would be ironic if the small percentage of people who survived a massive, widespread social collapse included many of those who invented and deployed the technologies that helped to cause it.

15.3.3 Populism to Fascism

Collapse is not the only dystopian outcome. I argued previously that self-driving vehicles may well bring with them the undeniable start of widespread lasting unemployment, and perhaps a consequent panic. In some countries at least, that may bring forth strong leaders who promise security, and law and order. In these countries, society may peer over the brink and decide collectively to step back and away from collapse. They may accept a trade-off in which they surrender much of their liberty, and perhaps many of the rights they used to take for granted, in exchange for a guarantee of some kind of rule of law.

Seizing control of the media, of social media, and making full use of all the cameras and sensors in the environment, a determined government could probably exercise a formidable degree of control, and could perhaps cow and crush any resistance. In the absence of a genuine solution to the challenges posed by technological unemployment, various bogus solutions would be tried out, and their abject failure would be denied and covered up. The result would almost inevitably be economic reversal and hardship, but again, this might be an acceptable alternative to collapse.

The populist leader would doubtless demand slavish obedience and the outward expression of loyalty. There would also be the demonisation of some kind of 'out' group, defined by its race, religion or political beliefs. There would also be the mobilisation of fear and hatred towards an external enemy, a rival nation or group of nations which required the domestic population to unite and make sacrifices to defeat it.

We have seen this story too many times before, in fascist regimes of the right and the left. It always creates misery, and it usually leads to war. So, in the long run, it may not be an alternative to collapse at all.

15.3.4 Fracture and Then Collapse

Another dystopian vision which could be realised some years further into the future is speciation: the gods and the useless scenario that we explored in Section 14.5. It could be a stable if unappetising arrangement, like the world envisioned in Huxley's *Brave New World*. But the encounters between the Spanish conquistadores and the Aztecs in Mexico and the Inca in Peru are two of many which suggest that encounters between two human civilisations do not work out well for the one which is less technologically advanced. It seems a reasonable expectation that sooner or later the gods and the useless scenario would lead to serious trouble for one or both parties.

15.4 PROTOPIA

15.4.1 Utopia, Dystopia, Protopia

Kevin Kelly is a writer, and the founding editor of *Wired* magazine. He has been called the most interesting man in the world.[5] I have no idea whether he enjoys the burden of that appellation, but he does produce a lot of interesting ideas. One of his good ones is protopia.

Too much of today's thinking about the future is dystopian, and that is partly because too many people fail to realise just how much progress *Homo sapiens* has made in the last few centuries and decades. It is natural and indeed helpful for our species to be discontented: if we weren't discontented, we probably wouldn't struggle to make the world a better place. But it can lead to dangerous misunderstandings.

Many people think that all politicians are corrupt, and that all corporations are run by Bond villains who are greedy and bent on world domination. Most of us could think of some group, clique or tribe that we are suspicious, fearful or disdainful of. But the truth is that in most of the world, today is the best time there has ever been to be alive. Most people in developed countries today live better than kings and queens did a couple of centuries ago. We live longer, eat better, have better healthcare, and inconceivably better access to information and entertainment than previous generations. If you doubt this, take a look at one or two of the late Hans Rosling's delightful and inspiring TED talks,[6] and browse the charts at 'Our World in Data' compiled by Oxford University researcher Max Roser.[7]

Of course, everything could go horribly wrong tomorrow. There might even be an iron law of nature that when civilisations reach a certain stage they either blow themselves up or create machines to do it for them. But from where we stand today, there is no reason to believe that. It seems more likely that the future is open, and potentially very good indeed.

Utopian visions of the future are less common, but they are also problematic. A future in which life has to all intents and purposes become perfect sounds sterile and boring. It is also highly improbable: the more we learn about the universe, the more we discover that we don't know, so it seems unlikely the universe will one day stop presenting us with puzzles and challenges. Perhaps this is why the two best-known literary descriptions of utopias, Thomas More's *Utopia* and Voltaire's *Candide*, are essentially critiques of the societies they lived in rather than recipes for an ideal future one.

So, it is refreshing to read Kelly saying this: 'I am a protopian, not a utopian. I believe in progress in an incremental way where every year it's better than the year before but not by very much – just a micro amount.

I don't believe in utopia where there's any kind of a world without problems brought on by technology. Every new technology creates almost as many problems that it solves'. But crucially, it gives us 'a choice that we did not have before, and that tips it very, very slightly in the category of the sum of good'.[8]

15.4.2 The Good Life

That 'sum of good' might be what the ancient Greeks called 'eudaimonia': the good life that we all seek (or at least, should seek), in which we humans flourish. The Greeks debated vigorously whether eudaimonia consisted of happiness, or virtue, or both. Their concept of virtue included being good at something as well as the moral virtues lauded by Christianity and other monotheistic faiths. At the risk of oversimplifying, Socrates, and later on the Stoics, thought that eudaimonia required the exercise of virtue alone, because virtue is both necessary and sufficient for happiness. Aristotle disagreed, saying that eudaimonia requires both happiness and virtue. A completely virtuous person would not enjoy eudaimonia if, for instance, all her children died. Epicurus added another twist, arguing that happiness was the only necessary ingredient of eudaimonia, but that happiness is impossible without leading a life that is also virtuous. Confused? Welcome to the human condition.

15.4.3 The Fourth Scenario

The first two scenarios explored in this chapter were perfectly acceptable, but, I suspect, unrealistic. The third was dystopian. What would the fourth scenario – protopia – look like? I think it might unfold in four stages: a plan, big welfare, Star Trek and collectivism. We will flesh these stages out in the next chapter.

NOTES

1. https://edge.org/conversation/john_markoff-the-next-wave.
2. https://en.wikipedia.org/wiki/Milgram_experiment.
3. http://www.prisonexp.org/.
4. http://www.newyorker.com/magazine/2017/01/30/doomsday-prep-for-the-super-rich.
5. http://fourhourworkweek.com/2014/08/29/kevin-kelly/.
6. https://www.ted.com/speakers/hans_rosling.
7. https://ourworldindata.org/.
8. https://www.edge.org/conversation/kevin_kelly-the-technium.

CHAPTER 16

Protopian Un-forecast

16.1 CAVEAT

16.1.1 Three Snapshots of a Positive Scenario

This chapter offers three snapshots of a possible future, taken in 2025, 2035 and 2045. Their purpose is to make the possibility of technological automation seem more real and less academic, and to explore how a positive outcome might unfold, with a concluding vision of an economy of radical abundance which has been achieved without massive social dislocation. In each section, there is a summary of the impact of cognitive automation on society, and a very brief description of the level of automation in a number of industries.

Before we start, there is an important caveat.

16.1.2 Unpredictable Yet Inevitable

We know that all forecasts are wrong. The only things we don't know are by how much and in what direction. The future generally turns out to be not only different to what we expect, but also much stranger. Cast your mind back to 2005. Pretty much everyone thought that mobile phones would continue to get smaller, and Facebook was limited to a few thousand universities and schools. Today, just a decade later, larger smartphones are a bit of a thing, and Facebook's valuation has overtaken that of Walmart, the world's largest shopkeeper.[1] Trying to predict how the world will look in 2031 is like trying to predict the weather on Saturday two months from now. There are just too many variables.

And yet in hindsight, what happens appears not only natural, but almost inevitable.

The smartphone is a good example. Pretty much nobody suggested 30 years ago that we would all have telephones in our pockets which would contain powerful artificial intelligences, and which would only occasionally be used for making phone calls. After all, at the time a mobile phone was a fairly hefty device, the size of a small dog. But now that it has happened, it seems obvious, logical and perhaps even inevitable.

Here's why. We humans are highly social animals, and our social habits are facilitated by language. Because we have language we can communicate complicated ideas, suggestions and instructions: we can work together in teams and organise; we can defend ourselves against lions and hostile tribes, we can hunt and kill mammoths, produce economic surpluses and develop technologies.

It is often said that no species is more savage and more violent than humans. This is no more true than the claim that Americans are more violent than other nationalities because their murder rate is higher than other developed countries. The only reason why humans kill more than other species is that we have more and better weapons.

Humans live cheek by jowl in cities containing millions. This is remarkable: no other carnivorous species can assemble more than a few dozen of its members in a limited space without them killing each other in rivalry for food, sex or social dominance. Other species lack our sophisticated ways to communicate and collaborate. Our bigger brains allow us to establish laws and cultural norms which govern the way we interact. We make up stories and agree to believe in them collectively, regardless of whether there is any evidence for them. These stories – about abstract concepts like gods and kingship, nations and ideologies, money and art – give us powerful reasons to cooperate and work together; even to die together.

Non-human primates spend hours every day grooming the other members of their tribe to reassure them that they will not sink their teeth and nails into them. It works, but it is inefficient, and means they cannot readily add new members to their tribe. Humans, by contrast, can walk past complete strangers on a crowded street without a second thought. Our superpowers are communication, and our capacity to sustain mutual belief in things which we either know to be illusory, or for which we have no evidence (like religion, monarchy, currency, democracy and nationality). It is thanks to these abilities that we control the fate of this planet and every species on it.

(This means that the old cliché that our dominance is based on our capacity for rational thought is – unlike most clichés – untrue.)

So, although it wasn't predicted in advance, in hindsight it is entirely logical that our most powerful technology, artificial intelligence (AI), would first become available to most of us in the form of a communication device.

The way the economic singularity unfolds will probably be like that. Our attempts to forecast the impact of technological unemployment – assuming it arrives – will probably look absurd in hindsight. But when we get there, the outcome will seem not only natural, but perhaps even inevitable.

16.1.3 Un-forecasting

I am labouring this point because I want to be clear. The description of a possible future that follows is not a prediction. The only thing I am confident of is that the future will not be like this.

Instead, this timeline is intended to serve two functions. First, it is a rhetorical device to make some of the seemingly outlandish ideas in this book more plausible. The arguments in Chapter 13: Is It Different This Time? that machines will automate our jobs away have been either abstract or fragmentary, and as such, some readers may find them implausible. I'm hoping that the timeline will help make the possible future of an economic singularity seem less academic, less theoretical and more real. And more hopeful.

Secondly, drawing up timelines like this one should help us to construct a valuable body of scenarios. Even when we know the future is unpredictable, it is still essential to make plans. There is good sense in the old cliché that failing to plan is planning to fail. If you have a plan, you may not achieve it, but if you have no plan, you most certainly won't.

In a complex environment, scenario development is a valuable part of the planning process. None of the scenarios will come true in their entirety, and many will be completely off the mark. But parts of some of them may approximate parts of the outcome. Thinking through how we would respond to a sufficient number of carefully thought-out scenarios should help us to react more quickly when we see the beginnings of what we believe to be a dangerous trend.

16.1.4 Super Un-forecasting

The art of constructing a useful scenario is similar to forecasting, which has been extensively studied by Canadian political scientist Philip Tetlock, co-author of the book *Superforecasting: The Art and Science of Prediction*. He has found that the best forecasters share a number of traits. First, they treat their views about what will happen as hypotheses, not firm beliefs. If the evidence changes, they change their hypothesis.

Secondly, they look for numerical data. Now we all know that there are lies, damned lies and statistics, and that data is often used in public debate in the same way that a drunk uses a lamp-post: more for support than for illumination. But used carefully and honestly, data is our friend. It is, after all, the root of the scientific revolution that has lifted most of our species out of poverty and squalor.

Thirdly, they look for context. He cites the example of guests at a wedding, admiring the beauty and grace of the bride and the dashing good looks of the bridegroom, and assuring each other that they will share a long and happy life together. The super-forecaster is a contrarian, noting that around half of all marriages fail, and that the failure rate increases with second and third marriages, especially when one or the other partner has a history of infidelity, as with the happy couple today. If she is a tactful super-forecaster, she keeps these thoughts to herself.

Ironically, super-forecasters are often not the people who get listened to in discussions about the future. We tend to pay more attention to those who speak most confidently, and offer clarity and certainty. People who equivocate and offer measured suggestions often don't cut through the noise.

So here goes, with the equivocation minimised.

16.2 2025: PANIC AVERTED

Vehicles without human drivers are becoming a common sight in cities and towns all over the world. Professional drivers are starting to be laid off, and it is clear to everyone that most of them will be redundant within a few years. At the same time, employment in call centres and the retail industry is hollowing out as increasingly sophisticated digital assistants are able to handle customer enquiries, and the move to online shopping accelerates. The picking function in warehouses has been cost-effectively automated, and we are starting to see factories which normally have no lights on because no humans are working there.

Many companies have laid off some workers, but most have reduced their headcount primarily by natural wastage: not replacing people who retired or moved on. As a result, there have been fewer headlines about massive redundancies than some people feared, but at the same time it has become much harder for people to find new jobs. The unemployment rate among new graduates is at historically high levels in many countries.

But instead of panicking, the populations of most countries are reassured by their political and business leaders telling them that they have a plan. This is possible because governments and other large organisations started to sponsor several prestigious new think tanks a few years before, in the late 2010s. Gradually, a consensus has formed within and between these groups that the Star Trek economy is indeed achievable, and numerous policy documents have been published explaining how it could be reached.

1. *Transportation*: The full automation of driving took a little longer than the most bullish pioneers expected, but not much. Regulation and popular resistance was less of a hurdle than some had feared,

and as soon as the technology was ready, it was implemented in almost every developed country, and soon afterwards in most of the developing world.

Elon Musk was right: the early buyers of self-driving cars are offsetting the purchase cost by renting them out through apps like Uber to other people when they are not using them. Many people have decided not to upgrade to a self-driving car, but instead rely on the fleets of automated taxis which are offered by partnerships between auto manufacturers and technology companies, as well as more established fleet enablers like Uber. The technology to retro-fit autonomy into older cars looks likely to become price competitive, and various experiments with car-sharing, car-pooling and so on are appearing.

The biggest impact is being felt in the commercial vehicle sector. Fleet managers are painfully aware that their competitors are cutting out the cost of their human drivers – or preparing to – and they know that they must do likewise or go out of business. The non-driving tasks carried out by drivers, such as checking their loads, protecting them from robbery, helping to load and unload, have turned out to be achievable through other means.

2. *Manufacturing*: Industrial robots are cheap enough, and easy enough to programme and maintain, that manufacturers normally choose to buy a robot rather than hire a human when they expand a line. The most sophisticated manufacturers of cars and electrical equipment now have a few 'lights-out' factories where no humans are normally required, but this is still rare. Likewise, although a few manufacturers have laid off large numbers of workers, most have yet to make this step, relying instead on natural wastage to reduce costs.

3. *Agriculture*: Farmers are experimenting with robots for both crops and animal husbandry. On a growing number of farms with high-value crops, small wheeled devices patrol rows of vegetables, inter-rogating plants which don't appear to be healthy specimens, and eliminating weeds with herbicides, or targeted jets of ecologically sound hot water. Cattle are entirely content to be milked by robots, so fewer members of the declining population of farm workers still have to get up before daybreak every day.

4. *Retail*: The shift towards purchasing goods and services online continues, and there is growing automation within shops. In many supermarkets, shoppers no longer have to unload and reload their trolleys: the goods are scanned while still inside their baskets. Fewer attendants are required in the checkout area. In fast food outlets, so-called 'McJobs' are disappearing as burgers and sandwiches are assembled and presented to customers without being handled by a human.

5. *Construction*: Property construction firms are increasingly experimenting with prefabricated units, but most construction projects remain subject to great variability of on-site conditions, including the foundations. Robots which can handle this unpredictability are still too expensive to replace human construction workers. There are experiments with exoskeletons for construction workers, but these are still expensive.

6. *Technology*: The tech giants are still fighting to recruit and retain machine learning experts; the salaries and bonuses offered were previously unknown outside financial services and professional sports. There has been lively debate about whether two of the cloud-based platforms should be broken up on monopoly grounds, but calls to do so have so far been denied because there was no evidence that the platforms were actually operating to the detriment of the consumer. The platforms were also able to establish that being broken up would destroy the linkages that made them so powerful, and so beneficial to consumers.

7. *Utilities*: Water companies and power generation and transmission firms are building fleets of tiny robots and drones which patrol pipes and transmission lines, looking for early warning signs of failure.

8. *Finance*: Retail banking is mostly automated and web-based, and consumer feedback on the quality of service is improving. Wealthy people now get some of their investment advice directly from automated systems, but human investment advisers still serve most of the market. In corporate finance, human advisers show no signs of being replaced, although their back office systems are heavily automated.

9. *Call centres*: Enquiry handling that was offshored to India and then repatriated to home countries is now being offshored again – this time to machines housed in cold climates where the cost of keeping the servers cool are lower.

10. *Media and the arts*: Virtual and augmented reality equipment is now very impressive, but still improving fast. Spending time in virtual reality (VR) arcades is one of the most popular pursuits for teenagers and young adults. But despite numerous false dawns, augmented reality (AR) and immersive VR for consumers is still not good enough for the mass market, and smart glasses remain a niche business for enterprise applications. As usual, porn and sport look like being the killer apps for consumer VR, but there are unexpected hits too, such as 'how-to' shows about parenting and relationship enhancement.

11. *Management*: Managers whose jobs consisted of processing information and passing it on are looking for (and struggling to find) new work, but many managers who deal with other people – staff or customers – are still employed.

12. *Professions*: The tedious jobs which traditionally provided training wheels for accountants and lawyers ('ticking and bashing' for auditors and 'disclosure' or 'discovery' for lawyers) are increasingly being handled by machines. Sceptics about technological unemployment point out that the amount of work carried out by professional firms has actually increased, as whole categories of previously uneconomic jobs have become possible, and that professionals are kept busy because the machines still need training on each new data set. But fewer trainees are being hired, and thoughtful practitioners are writing articles in their trade magazines asking where tomorrow's qualified lawyers and accountants will come from.

13. *Medical*: AIs are ingesting data sent by patients from their smartphones and carrying out triage. Sometimes they respond with simple diagnoses and treatment recommendations, sometimes they pass the enquiry to a human doctor. Medical professionals and regulators are nervous about these experiments, but there is growing evidence of positive outcomes. Overall, the workload for the medical profession has increased rather than reduced, as patients are taking a more active and better informed approach to their own healthcare.

Hospitals in Japan are using robot nurses to great effect, and these trials are followed with great interest elsewhere. Pharmaceuticals designed to raise the IQ of adults are in clinical trials.

14. *Education*: There is now overwhelming evidence that techniques such as flipped learning and competency-based learning produce impressive outcomes, but they are far from being universally adopted. Competitive environments, such as the United Kingdom's private school system, are experimenting with AI-assisted learning, in which students have personal AI tutors. They are rudimentary at present.

15. *Government*: There is a worldwide drive to get most government 'services' delivered online and cheaper.

16.3 2035: TRANSITION

Large numbers of people are now unemployed, and welfare systems are much enlarged almost everywhere. Pretty much everyone now accepts that universal basic income is not the solution, because there is no point paying generous welfare to the many people who remained in lucrative employment, and because it is widely agreed that a basic income is insufficient for the rest. The new welfare has many incarnations, and interesting new experiments are springing up all the time. Concepts like progressive comfortable income (PCI) and human elective leisure programme (HELP) are being discussed not only in the think tanks, but in kitchens, bars and restaurants all over the world.

1. *Transportation*: Most jurisdictions now have roads which are off-limits to human drivers. Insurance premiums have plummeted, and fears about self-driving cars being routinely hacked have not been realised. A vocal minority of citizens (which, to general surprise, comprises equal numbers of men and women) are scathing about this arrangement, dubbing the communal cars THEMs, ('tedious horizontal elevator machines').

   Professional drivers are now extremely rare. Their disappearance was resisted for a while – sometimes fiercely. Autonomous vehicles were frequently attacked in some places, a favourite tactic being to spray-paint the cameras and LIDARs (Light Detection and Ranging) they rely on. Some high-profile arrests and jail sentences quickly put a stop to the practice.

   Deliveries of fast food and small parcels in major cities are now mostly carried out by autonomous drones, operating within their own designated level of airspace. Sometimes the last mile of a delivery is carried out by autonomous wheeled containers. For a while, teenagers delighted in 'bot tipping', but with all the cameras and other sensory equipment protecting the bots, the risk of detection and punishment became too high.

   Many years after they first became an emblem of the future, flying cars finally became a real phenomenon. They fly through designated air corridors between major cities, generally taking off from and landing on the roofs of tall buildings. They are mercifully quiet, powered by multiple electric-powered rotors.

2. *Manufacturing*: Many large factories and warehouses are now dark: no light is required because no humans work there. People are becoming a rarity in smaller sites too.

   3D printing has advanced less quickly than many expected, as it remained more expensive than mass production. But it is common in niche applications, like urgently required parts, components with complex designs and situations where products are bespoke, as in parts of the construction industry. They have an impact on businesses and the economy far greater than their modest output level would suggest.

3. *Agriculture*: Farmers are moving heavily into leisure services, as their families and staff are losing their roles to robots.

4. *Retail*: Online shopping reaches 75% of all retail purchases, with a small but growing number of items being 3D printed domestically or in neighbourhood facilities, often with an element of customisation by the consumer. In the high street shops that remain, human shop assistants are starting to be replaced by robots, except in high-margin sectors where humans help to create an experience rather than facilitating transactions.

5. *Construction*: Human supervision is still the norm for laying foundations, but prefabricated (often 3D-printed) walls, roofs and whole building units are becoming common. Robot labour and humans in exoskeletons are increasingly used to assemble them. Drones populate the air above construction sites, tracking progress and enabling real-time adjustments to plans and activities.

6. *Technology*: Following the popularity of 'wearables', electronic devices that are worn on the body as accessories, or woven into clothing, the first 'insideables' are appearing, and have been made fashionable by Lord Beckham, the ex-footballer and fashion magnate. The IoT has materialised, with everyone receiving messages continuously from thousands of sensors and devices implanted in vehicles, roads, trees, buildings and so on. Fortunately, the messages are intermediated by personal digital assistants, which have acquired the generic name of 'Friends', but whose owners often endow them with pet names.

New types of relationship and etiquette are evolving to govern how people interact with their own and other peoples' Friends, and what personalities the Friends should present. Brand loyalty to the companies which provide the best Friends software is fierce.

There is lively debate about the best ways to communicate with Friends and other computers. Concealed microphones handle much of the traffic, but millions of people are also learning how to use one-handed keyboards which liberates them from traditional keyboards at times when voice is inappropriate. Some believe these new keyboards will be quickly superseded by brain–computer interfaces (BCI), but this has made less progress than its early enthusiasts expected.

Another promising technology is tattoos worn on the face and around the throat which have microsensors to detect and interpret the tiny movements when people sub-vocalise, that is, speak without actually making a noise. The tattoos are usually invisible, but some people have visible ones which give them a cyborg appearance.

A growing amount of entertainment and personal interaction is mediated through VR. Good immersive VR equipment is now found in most homes, and it is increasingly rare to see an adolescent in public outside school hours.

Polls suggest that most people now think that artificial general intelligence (AGI – machines which equal or surpass human cognition in all domains) is a serious possibility within a generation or two. Significant expenditure is flowing into research on how to make sure the outcome is positive, and the moral and religious implications are hotly debated.

7. *Utilities*: In many organisations, most operations are now automated. The main role of humans in these organisations is testing security arrangements. Several hundred people died in two significant hacking incidents – one in the United States and one in Europe. This has prompted huge investment in upgraded security arrangements. In another high-profile incident, an AI system managed the disaster containment and recovery process flawlessly, and much faster than humans could have.

8. *Finance*: Retail banking is now fully automated, and investment advice is going the same way. Corporate financiers are in retreat, and their previously stratospheric incomes have fallen sharply.

9. *Call centres*: Almost no humans now work in call centres.

10. *Media and the arts*: All major movies made by Hollywood and Bollywood are now produced in VR, along with all major video games. To general surprise, levels of literacy – and indeed book sales – have not fallen. In a number of genre categories, especially romance and crime, the most popular books are written by AIs.

Major sporting competitions have three strands: robots, augmented humans, and un-augmented humans. Audiences for the latter category are dwindling.

Long-distance communication is massively improved by VR Skype.

Dating sites have become surprisingly effective. They analyse videos of their users, and allocate them to 'types' in order to match them better. They also require their members to provide clothing samples from which they extract data about their smells and their pheromones. The discovery that relationship outcomes can be predicted with surprising accuracy with these kinds of data has slashed divorce rates.

11. *Management*: The ranks of middle management are thinning out. Shareholders are investing heavily in distributed autonomous corporations (DACs), firms consisting of unsupervised AIs which create new business models and strategies and transact with other firms without any humans in the loop.

12. *Professions*: Partners in law firms and accountancy firms are working shorter hours. Human intakes to these firms are dwindling. Most criminal law cases now relate to digital crime: with tiny, powerful cameras everywhere accompanied by highly effective facial recognition technology, the few remaining perpetrators of physical crime are generally caught quickly.

13. *Medical*: Opposition to the smartphone medical revolution has subsided in most countries, and most people obtain diagnoses and routine health check-ups from their Friends several times a week. Automated nurses are becoming increasingly popular, especially in elder care.

   Several powerful genetic manipulation technologies are now proved beyond reasonable doubt to be effective, but backed by public unease, regulators continue to hold up their deployment. Cognitive enhancement pharmaceuticals are available in some countries under highly regulated circumstances, but are proving less effective than expected. There are persistent rumours that they are deliberately being engineered that way.

   Ageing is coming to be seen as an enemy which can be defeated.

14. *Education*: Data on learning outcomes is steamrollering teachers' resistance to new approaches. Customised learning plans based on continuous data analysis are becoming the norm. Teachers are becoming coaches and mentors rather than instructors. More and more schools are experimenting with classroom AIs.

15. *Government*: There is growing pressure to reduce the numbers of politicians and civil servants, as more and more government services are automated. Many jurisdictions are debating the merits of using technology to enable direct democracy, which is being pioneered by Switzerland. Most people are sceptical, fearing the tyranny of the temporary majority. Policemen in most countries record all interactions with members of the public, and public satisfaction levels with them are generally rising.

16. *Charities*: Non-profit organisations are enjoying a surge, thanks to an influx of talent as capable people can't find work elsewhere.

16.4 2045: THE STAR TREK ECONOMY MATERIALISES

AI systems have driven the cost of production of most non-luxury goods and services close to zero. Few people pay more than a token amount for entertainment or information services, which means that education and world-class healthcare are also much improved in quality and universally available. The cost of energy is dramatically reduced also, as solar power can now be harvested, stored and transmitted almost for free. Transportation involves almost no human labour, so with energy costs so low, people can travel pretty much wherever they want, whenever they want. The impressive environments available in VR do a great deal to offset the demand for travel that this might otherwise have created.

Food production is almost entirely automated, and the use of land for agriculture is astonishingly efficient. Vertical farms play an important role in high-density areas, and wastage is hugely reduced. The quality of the housing stock, appliances and furniture is being continuously upgraded. A good standard of accommodation is guaranteed to all citizens in most developed countries, although of course there are always complaints about the time it takes to arrive. Personalised or more luxurious versions are available at very reasonable prices to those still earning extra income.

Almost no one in developed countries now lives in cramped, damp, squalid or noisy conditions. Elsewhere in the world, conditions are catching up fast.

Other physical goods like clothes, jewellery and other personal accessories, equipment for hobbies and sports and a bewildering array of electronic equipment are all available at astonishingly low cost. The Star Trek economy is almost mature. But access to goods and some services is still rationed by price. Nobody – in the developed world at least – wants for the necessities of civilised life, but nobody who is not employed can afford absolutely everything they might wish for. It is generally accepted that this is actually a good thing, as it means the market remains the mechanism for determining what goods are produced, and when.

Unemployment has passed 75% in most developed countries. Among those still working, nobody hates their job: people only do jobs that they enjoy. Everyone else receives an income from the state, and there is no stigma attached to being unemployed, or partially employed. In most countries, the citizens' income is funded by taxes levied on the minority of wealthy people who own most of the productive capital in the economy, and in particular on those who own the AI infrastructure. The income is sufficient to afford a very high standard of living, with access to almost all digital goods being free, and most physical goods being extremely inexpensive.

In many countries, some of the wealthy people have agreed to transfer the productive assets into communal ownership, either controlled by the state or by decentralised networks operated using blockchain technology. Those who do this enjoy the sort of popularity previously reserved for film and sports stars.

Some countries mandated these transfers early on by effectively nationalising the assets within their legislative reach, but most retreated from this approach when they realised that their economies were stagnating, as many of their most innovative and energetic people emigrated. Worldwide, the idea is gaining ground that private ownership of key productive assets is distasteful. Most people do not see it as morally wrong, and don't want it to be made illegal, but it is often likened to smoking in the presence of non-smokers. This applies particularly to the ownership of facilities which manufacture basic human needs, like food and clothing, and to the ownership of organisations which develop the most essential technology – the technology which adds most of the value in every industry sector: AI.

The gap in income and wealth between rich and poor countries has closed dramatically. This did not happen because of a transfer of assets from the West to the rest, but thanks to the adoption of effective economic policies, the eradication of corruption, and the benign impact of technology in the poorer countries.

Another concern which has been allayed is that life without work would deprive the majority of people of a sense of meaning in their lives. Just as amateur artists were always happy to paint despite knowing that they could never equal the output of a Vermeer, so people now are happy to play sport, write books, give lectures and design buildings even though they know that an AI could do any of those things better than them.

Not everyone is at ease in this brave new world, however. Around 10–15% of the population in most countries suffers from a profound sense of frustration and loss, and either succumbs to drugs or indulges almost permanently in escapist VR entertainment. A wide range of experiments is under way around the world, finding ways to help these people join their friends and families in less destructive or limiting lifestyles. Huge numbers of people outside that 10–15% have occasional recourse to therapy services when they feel their lives becoming slightly aimless.

Governments and voters in a few countries resisted the economic singularity, seeing it as a dehumanising surrender to machine rule. Although they found economically viable alternatives at first, their citizens' standard of living quickly fell far behind. Several of these governments have now collapsed like the communist regimes of Eastern Europe in the early 1990s, and the rumours that President-for-life Putin met a very grisly end in 2036 have never been disproved. The other hold-outs look set to follow – hopefully without violence.

1. *Transportation*: Humans do not drive vehicles on public roads, and very few commercial vehicles have human attendants. Young people do not take driving tests. Humans do drive vehicles for sport, but even motor sports are now mostly competitions between self-driving cars. Many people now live nomadic lifestyles in RVs. They spend a few days in one place and then, often overnight, their automated RVs drive to them to their next destination. The choice of destination is often governed by who else is going to be there, or a special event that is of interest. Selecting the next destination and arranging who to meet there is usually mediated by AIs, as is the route planning and driving.

2. *Manufacturing*: Almost all factories and warehouses are dark. 3D printing still accounts for a minority of the total goods manufactured, but it is becoming competitive with some forms of mass production.

3. *Agriculture*: Robots do most farm work.

   Some countries have large communally owned agricultural processing concerns which send out meals on drones in a service often described as Netflix for food.

4. *Retail*: The great majority of items are now bought online. Retail outlets on high streets and city centres are mostly experiential rather than transactional, and mostly staffed by AIs and robots.

5. *Construction*: Robots now carry out most of the work on construction sites.

6. *Technology*: Since AI provides a large proportion of the value in most products and services, there is a major concentration of capital and wealth in the hands of shareholders and key employees in this sector. Its foremost talent is now applied to developing AGI and making sure that it is safe for humans. The Internet of things (IoT) is all-pervasive, and the environment appears intelligent.

   The companies that provide Friends have been obliged to make them open-source. Friends are so critical to everyone's lives that being restricted to any one company's walled garden was unacceptable.

7. *Utilities*: Overwhelmingly automated.

8. *Finance*: Overwhelmingly automated.

9. *Call centres*: Unchanged.

10. *Media and the arts*: In sports, robot competitions now generate larger audiences than their human counterparts. The International Olympics Committee de-lists the human versions of around half of all sports.

    Haptic body suits combined with VR headsets now provide truly immersive virtual environments. Counselling (by AIs) is required by a section of the population who struggle to maintain the distinction in their minds between reality and VR.

To general surprise, people still read books, but they are very different products now, with holographic illustrations, and often with several alternative story lines developed by their AI authors, which readers can choose between.

Dating sites are now mostly accessed by personal digital assistants. 'My Friend likes your Friend' was a standard conversation opener for a while.

11. *Management*: Many companies now consist of just a few strategists, whose main role is to forecast the optimal business model for the next financial quarter, but they are struggling to keep up with their AI advisers.

12. *Professions*: Accountancy and the law are largely automated.

13. *Medical*: Demand for human doctors is dwindling and professional nursing has been almost entirely automated. Everyone in developed economies has their health monitored continuously by their Friends. Most people spend a certain amount of time each week visiting family, friends and neighbours who are unwell, just to converse. Sick and disabled people are greatly comforted by their relationships with talking AI companions, some resembling humans, others resembling animals.

Significant funds are now allocated to radical age extension research, and there is talk of 'longevity escape velocity' being within reach – the point when each year, science adds a year to your life expectancy. Most forms of disability are now offset by implants and exoskeletons, and cognitive enhancements through pharmaceuticals and BCI techniques are showing considerable promise.

14. *Education*: The sector has ballooned, with many people now regarding it as recreation rather than work. Most education is provided by AIs.

15. *Government*: Safeguards have now been found to enable direct democracy to be implemented in many areas. Professional politicians are now rare.

NOTE

1. http://money.cnn.com/2015/06/23/investing/facebook-walmart-market-value/.

CHAPTER 17

The Economic Singularity

Summary and Conclusions

17.1 SUMMARY OF THE ARGUMENT

17.1.1 Automation and Unemployment

We cannot be certain, but it looks likely that improvements in machine intelligence over the next few decades are going to make it impossible for most humans to earn a living. We would be wise to devote some resource to working out how to deal with this development in case it does happen – indeed we would be foolish not to.

During the Industrial Revolution, concerns that automation would lead to permanent mass unemployment turned out to be unfounded – unless you were a horse. (The Engels pause was lengthy, but not permanent.) Instead, automation raised productivity and output across the economy. The unfounded concerns became known as the Luddite fallacy.

In the information revolution, mankind's fourth great wave of transformation (although definitely not the fourth industrial revolution), machines are increasingly able to outperform humans in cognitive tasks. This is likely to put humans in the predicament that horses were placed in by the Industrial Revolution.

In 1900, 40% of American workers were employed in agriculture, and that has now fallen below 3%. The farm workers found better jobs elsewhere in the economy, sometimes in occupations which their parents could not have imagined. But horses didn't. The year 1915 was 'peak horse', with most of the 21.5 million horses in America working on farms; now there are virtually none. The difference between the horses and the humans is that when machines took over the muscle jobs, humans had something else to offer: our cognitive, emotional and social abilities. Horses had nothing else to offer, and their population collapsed to a mere two million.

Past rounds of automation have mostly been mechanisation, the replacement of human and animal muscle power. The coming wave of automation will substitute our cognitive abilities. Machines don't need to become artificial general intelligences (AGIs) to displace most of us from our jobs. They simply have to become better than us at what we do for a living. They are close to or at parity with us with regard to many forms of pattern recognition, including image recognition and speech recognition. They will soon be considerably better than us, and continuing to improve at an exponential rate. They are catching up on our ability to process natural language, and they will overtake us there as well.

Once a machine can do your job, it will quickly be able to do it faster, better and cheaper than you can. Machines don't eat, sleep, get drunk, tired or cranky. And unlike human brains, their abilities will continue to improve at an exponential rate.

17.1.2 Sceptics

Sceptics offer two arguments to prove that automation cannot cause lasting widespread unemployment. Neither is impressive.

First, they claim that since automation has not caused lasting mass unemployment in the past, it cannot do so in the future. When stated this baldly, the weakness of this argument is plain: past performance is no guarantee of future outcome. If it was, we would never have learned how to fly. The observation that past rounds of automation have not caused lasting mass unemployment is not even true – if you consider the horse.

Secondly, some sceptics claim that since human wants and needs will never be fully satisfied, there will always be jobs for humans. This is also a weak argument: our needs may never be fully satisfied, but that does not mean humans will be able to compete with the machines in the attempt to do so.

In addition to these two arguments, sceptics offer some explanations of the kind of work we will all be doing when machines have taken over many or most of our existing jobs.

First, they suggest that we will race with the machines instead of against them, becoming 'centaurs', and occupying ourselves with the 'icebergs' of new work which machines have made possible. But these are only likely to be temporary respites, as the machines continue their rapid improvement.

Then they suggest that we will all do the caring jobs which machines cannot do because they have no consciousness and therefore no empathy. This turns out to be unconvincing too: people actually like being looked after by machines. They also suggest that we will buy goods and services from humans rather than machines out of chauvinism, and that we will all be artisans or entrepreneurs. None of these ideas stand up to scrutiny.

Finally, they invoke the magic job drawer, out of which will fly all sorts of jobs which we cannot imagine today because the technologies to make them possible have not been invented yet. They claim this is what happened during the Industrial Revolution: an agricultural labourer in 1800 could not have imagined that his descendant would become a web experience designer. But the idea that we all do jobs today that our grandparents could not have imagined is a myth. For example, 90% of all people working in America today are doing jobs which existed in 1990. And even if we do invent all sorts of new jobs which are unimaginable today, whatever new jobs we do invent, the machines will probably take over most of them as well. In the medium term.

In short, it is highly likely that within a generation or so, a large minority of people – perhaps the majority – will not be able to earn a living through jobs.

It is true that many well-informed people are sceptical of the technological unemployment thesis. They include economists like David Autor,

David Dorn and Robert Gordon, as well as technology industry leaders like Eric Schmidt, Sundar Pichai and Marc Andreessen. But arguments from authority should always be viewed with suspicion, and there are plenty of equally well-informed people on the other side of the debate, including artificial intelligence (AI) researchers like Stuart Russell and Andrew Ng, and technology industry leaders like Mark Zuckerberg, Elon Musk and Sam Altman.

17.1.3 The Upside

Fortunately, technological unemployment, if it happens, does not have to be bad news; in fact, it should be extremely good news. Some people are lucky enough to love their jobs and find fulfilment in them. For many more people, work is simply a way to generate an income for themselves and their families. It may provide a purpose, but it does not provide meaning. A world in which machines do all the boring work could be wonderful. They could be so efficient that goods and services could be plentiful, and in many cases free. Humans could get on with the important business of playing, relaxing, socialising, learning and exploring. Surely, this is what we should be aiming for.

People who believe that technological unemployment can lead to this world of freedom are the true optimists. People who believe that humans will have to remain in paid employment forever are the true pessimists.

17.1.4 Challenges

Reaching this positive outcome requires us to confront the following five challenges: meaning, economic contraction, income, allocation and cohesion. Although we probably have a considerable amount of time before these challenges bite hard in practice, the possibility of a panic erupting when the majority of people start to anticipate the likely changes means that the problem is more urgent than it might at first appear.

When people first take seriously the idea of the economic singularity, they often fear that unemployed humans will find their new lives hollow, lacking in meaning and perhaps even boring. This fear is probably mostly misgiven. For centuries, aristocrats in most countries didn't work for a living, and in many societies they viewed work as a demeaning activity to be avoided by 'people of quality'. Some of them got into trouble with drink, drugs and gambling, but only a small minority. Most of them seem to have led contented lives, however questionable we might find the economic systems they operated in.

Likewise, retirement is rarely considered a disaster in developed countries – as long as you retire with a sufficient income. Even though most of us only get to enjoy it when we are past our prime, most retirees find enough projects and pastimes to keep themselves busy and at peace. Numerous surveys have found that happiness is U-shaped: we are at our most content during childhood and retirement, and it is probably no coincidence that these are the periods in our lives when we don't work for a living. If we retired when still in our prime, we would be even better equipped to enjoy our lives of leisure.

The first really big challenge posed by the economic singularity is how to ensure that everyone has a good income – or rather, full access to the goods and services that are needed for a good life. (This overshadows and subsumes the problem of economic contraction.)

Many people think the answer is universal basic income (UBI), but in an economy where goods and services remain expensive, UBI will either be set too low to be sufficient, or it will be unaffordable. It cannot be paid for by soaking the rich, by eradicating bureaucracy or by taxing robots. Two of its major weaknesses are clearly stated in its very name: a universal payment wastes enormous sums on people who don't need it, and a basic income is inadequate and would not create a stable, sustainable society.

The effective way to solve the income problem is to drive close to zero the costs of all the goods and services required for a good life. In other words, to achieve the economy of radical abundance, also known as the Star Trek economy. Although this seems implausible at first sight, there are signs that it is in fact achievable.

If we do solve the income problem, there will still remain a problem of allocation: in a world where most people cannot vary their income, how do we decide who enjoys the scarce things in life, like the desirable detached houses in the better neighbourhoods, the penthouse apartments, the beach-front properties, the original Vermeers and the Aston Martins? Virtual reality might provide at least some of the answer.

The final challenge raised by the economic singularity may turn out to be cohesion. Capitalism, including the institution of private property, and a sensibly regulated market economy has been enormously beneficial for humanity. Along with the enlightenment and the consequent scientific revolution, capitalism has made our time the best era to be born human, bar none.

But a world where only a minority have jobs, and an elite owns the intelligent machines, may be a world of fantastic and entrenched inequality. Inequality is often overestimated as a contemporary social evil, but this post-economic singularity world will also be one where advancing technology makes available radical enhancements to our physical and cognitive performance. These enhancements will come along faster and faster, and groups with privileged access to them may start to diverge from everyone else and become a separate species. The author Yuval Harari has referred to this scenario in the chilling phrase, 'the gods and the useless'. The 'brave new world' depicted by Aldous Huxley in 1931 might be one of the least bad outcomes of this scenario.

If the post-economic singularity world needs a different type of economy, then we need to start thinking now about what that might be – and also how to get there. The damage that could be caused by an uneven or violent transition to the new world could be immense.

17.1.5 Scenarios

We have considered four potential outcomes to this process: no change, full employment, social collapse and the Star Trek economy.

Robert Gordon and John Markoff claim to believe that the great age of innovation is behind us, and technological progress today is just twiddling with unimportant apps. Estimable though these writers are, this is obvious nonsense. Google Search plus Wikipedia has given us something which our ancestors would have mistaken for omniscience, and it is clear to almost everyone that the advent of smarter and smarter machines is going to have enormous impacts on all of us.

Full employment is a more plausible scenario, but the case that it will certainly be true has not been made. It is hard to see why, given their continued exponential improvement, machines will not be able to replace humans in most paid work roles: it is just a matter of time.

The scenario of social collapse, or of societies falling prey to some kind of totalitarian control, is all too plausible. But it is obviously not acceptable: we must make sure it does not happen.

Which leaves the Star Trek economy, the economy of radical abundance. If it can be achieved – and achieved in time–it offers the potential for humans to flourish in ways that our ancestors could barely have dreamed of. It seems to be the best solution to the challenges posed by the economic singularity, and it is what we should be aiming for.

17.2 WHAT IS TO BE DONE?

17.2.1 We Need a Plan

At a conference in June 2017,[1] Professor Stuart Russell suggested that we should lock a group of economists and science fiction authors into a room and not let them out until they have come up with a plan for how to respond to technological unemployment. It's a terrific idea. We need a plan for how to navigate the economic singularity, and devising the plan will require both intellectual rigour and freewheeling creativity and imagination.

Scarcely a day goes by without an article appearing in the media debating whether robots will steal our jobs. Often, they are impressionistic, poorly argued pieces – it is depressing, for example, how often the reverse Luddite fallacy rears its ugly head. There are worryingly few places where sustained effort is being applied to solving the problem that these articles discuss. The best-known is the Oxford Martin Programme on Technology and Employment,[2] led by Dr Carl Benedikt Frey and Michael Osborne, who wrote the influential 2013 paper which suggested that 47% of American jobs would be displaced.

As we saw in Chapter 10: Ensuring that Superintelligence is Friendly, there are at least four permanently established organisations studying the risk to humanity posed by the potential arrival of superintelligence.[3] This is a very good thing, and the technological singularity is an existential threat to humanity, which the economic singularity is probably not. But the economic singularity is highly likely to arrive considerably sooner, and seeking solutions to the challenges it poses should be a priority.

We need to establish research institutes and think tanks to study the issues, to draw in diverse opinions, to canvas ideas and challenge them. And we need to start doing this now. The announcement in September 2017 that a UN agency is setting up a centre in the Hague to monitor developments in AI intelligence may be a constructive step.[4] The establishment by Silicon Valley luminaries Pierre Omidyar and Reid Hoffman (founders of eBay and LinkedIn respectively) of a $27 million fund for research into the impact of AI may be another.[5]

Cognitive automation has not yet begun to cause technological unemployment – or at least, not to any significant extent. The machine learning 'big bang' only happened a few years ago, and the enormous power of deep learning has yet to be harnessed by many organisations outside the tech giants. We have time to solve this problem, but we do not have time to waste. We should be monitoring developments and drawing up forecasts and scenarios. In Chapter 16: Protopian Un-forecast, we discussed how hard it is to make accurate forecasts, but failing to keep a lookout for approaching dangers (and opportunities) is foolish.

The view that most people will be rendered unemployable by machine intelligence within the next few decades is probably a minority opinion at the moment, but many of the people best-placed to understand what is coming do see it as a likely future.

Of course, there are many other causes whose proponents would make the same claim on the limited resources at our disposal: climate change, inequality and ocean pollution to name just three. Even the largest and wealthiest nations cannot expend sufficient resources to de-risk every danger that has been identified by somebody somewhere as important. We have to prioritise. I hope this book has persuaded you that the economic singularity should be high on that list of priorities.

17.2.2 Monitoring and Forecasting

One of the core tasks of these think tanks and research institutions would be to monitor developments within the major economies of the world, and alert us all when significant trends become apparent. But working empirically with data is not enough. The challenges we will face are in the future, not the past: there simply is no data about them yet. We need to forecast as well as monitor, in spite of the well-known problems with forecasting.

Prediction markets are probably part of the answer. People make their best estimates when they have some skin in the forecasting game. Offering people the opportunity to bet real money on when they see their own jobs or other peoples' jobs being automated may be an effective way to improve our forecasting. It also harnesses the power of a remarkable force which has attracted a lot of attention lately: the wisdom of the crowd.

In a prediction market, someone asks a question, often one requiring a yes/no answer, such as, 'Will Trump be impeached before the end of his first term?' Other people can buy and sell contracts for yes or contracts for no. The yes contracts are paid if the answer to the question turns out to be yes, and the no contracts are paid if the answer is no. The price of the contracts move, determined by supply and demand, and they are a prediction of the probability of the event occurring. So if the yes contract is priced in the market at 80, that means the market thinks there is an 80% chance of Trump being impeached before the end of his first term. The market ends when Trump's first term ends or when he is impeached, whichever happens first. If he is impeached, the price of the contract goes to 100, otherwise it goes to zero.[6]

17.2.3 Scenario Planning

Obviously, we are lacking sufficient information to draw up detailed plans for the way we would like our economies and societies to evolve. But we can and should be doing detailed scenario planning.

Scenario planning has been practised by military leaders since time immemorial. It was given the name by Herman Kahn, who wrote narratives about possible future for the US military while working for the RAND Corporation in the 1950s. (His suggestion that a nuclear war might be both winnable and survivable made him one of the inspirations for Dr Strangelove in the classic 1964 movie.[7]) Scenario planning was adopted by Shell after it (along with the rest of the oil industry) was disastrously wrong-footed by the rise of the oil cartel OPEC in the 1970s.[8]

Scenario planning is more art than science, but it can be a valuable discipline. When we commit our thoughts about a possible future to paper we are forced to consider them rigorously. Think tanks and research institutes consisting of smart people doing this work could make a valuable contribution.

An informal version of this is the daily business of futurists and futurologists, people who are often viewed with scepticism by the wider public. Perhaps that will change – in fact, perhaps futurology will come to be seen as a mission-critical profession. Science fiction writers and film makers also have an important role in providing vivid metaphors and warnings. It would be helpful if they (especially the film makers) explored more optimistic scenarios instead of revisiting the same old dystopian ones. It is a great shame, for instance, that no films have been made of Iain M Banks' hugely enjoyable *Culture* series books, in which humanity has prospered mightily.

17.2.4 Global Problem, Global Solution

Technological unemployment will affect every country, albeit at different rates and perhaps in different ways. We are in a race against time to find solutions to the problems it raises, but not in a race against each other. We can and should cooperate. At the moment, on the rare occasions when politicians talk about AI, they usually talk about making sure their country is in the lead in some kind of an AI race, or at least not falling too far behind. Certainly, it makes sense for every government to seek to promote the domestic development of AI, as that will benefit their people economically. But when it comes to the big issues – the singularities – they should be working together, sharing ideas, learning from each other.

It is likely that the territories where AI is being most vigorously developed will take the lead in much of this debate. Gavin Newsom, the front runner in the 2018 election for California's next governor, described the prospect of technological unemployment as 'code red, a fire-hose, a tsunami, that's coming our way', and admitted that 'I'm struggling to figure it out … I don't have the damn answer'.[9] He is not ruling out UBI or taxing robots as parts of the solution, and thinks education will be important too. Attentive readers will know that I have some issues with those prescriptions, but at least he is asking the questions. In August 2017, a member of San Francisco's city council launched a campaign called the Jobs of the Future Fund to study a state-wide 'payroll' tax on job-stealing machines.[10] Again, probably not the right answer, but at the moment it is encouraging just to see a politician asking the questions.

Politicians from other countries should be reaching out to these people to see what they have learned, and to see what they can in turn contribute to the discussion and the search for solutions.

17.2.5 Upsides, Downsides and Timing

The technological unemployment thesis might turn out to be false. Scenario two from Chapter 15: Four Scenarios, might turn out to be the reality, with humans remaining in employment for decades to come. If we allocate the resources to establish a few think tanks and research institutes, and the problem they were designed to tackle never arises, what have we lost? A few tens of millions of dollars at the most.

If, on the other hand, the technological unemployment thesis turns out to be correct, but we do nothing to prepare, what will we lose in that case? The panic described in Chapter 14: The Challenges, would be almost inevitable, leading to either an extreme version of populism and probably some kind of dictatorship, or perhaps directly to widespread social breakdown.

The sceptics about the technological unemployment thesis might be right: scenario two is not impossible. But they have no compelling arguments, and they may very well be wrong. Just hoping that they are right does not look like a wise policy.

The matter is urgent. If the panic is coming, it is likely to arrive within the next decade. The time to act is now.

NOTES

1. http://lcfi.ac.uk/media/uploads/files/CFI_2017_programme.pdf.
2. http://www.oxfordmartin.ox.ac.uk/research/programmes/tech-employment.
3. The Machine Intelligence Research Institute (MIRI) in Northern California, The Future of Humanity Institute (FHI) and the Centre for the Study of Existential Risk (CSER) in England's Oxford and Cambridge respectively, and the Future of Life Institute (FLI) in Massachusetts.
4. http://www.unicri.it/in_focus/on/UNICRI_Centre_Artificial_Robotics.
5. https://techcrunch.com/2017/01/10/omidyar-hoffman-create-27m-re-search-fund-for-ai-in-the-public-interest/.
6. There is a lot of useful insight into prediction markets here: https://bitedge.com/blog/prediction-markets-are-about-to-be-a-big-deal/.
7. Paul Boyer, *Dr Strangelove*, a chapter in *Past Imperfect: History According to the Movies*, edited by Mark C. Carnes.
8. http://s05.static-shell.com/content/dam/shell/static/public/downloads/bro-chures/corporate-pkg/scenarios/explorers-guide.pdf.
9. https://www.theguardian.com/us-news/2017/jun/05/gavin-newsom-governor-election-silicon-valley-robots.
10. https://www.wired.com/story/tax-the-rich-and-the-robots-californias-thinking-about-it/?mbid=social_twitter_onsiteshare.

Outroduction

SUMMARY: THE TWO SINGULARITIES

The technological singularity is the moment when (and if) we create an artificial general intelligence (AGI) which continues to improve its cognitive performance and becomes a superintelligence. Ensuring that we survive that event is probably the single most important task facing the next generation or two of humans – along with making sure we don't blow ourselves up with nuclear weapons or unleash a pathogen which kills everyone.

If we secure the good outcome to the technological singularity, the future of humanity is glorious almost beyond imagination. As DeepMind co-founder Demis Hassabis likes to say, humanity's plan for the future should consist of two steps: first, solve AGI, and second, use that to solve everything else. 'Everything else' includes poverty, illness, war and even death itself.

Well before we reach the technological singularity we will experience the economic singularity. The stakes here are not so high. If we find ourselves in the 'gods and the useless' scenario, or if our societies collapse as we fail to transition from modern capitalism to something more suitable for the new world, it is unlikely that every human will die. (Not impossible, though, as someone might initiate a catastrophic nuclear war.) Civilisation would presumably regress, perhaps drastically, but our species would survive to try again. Trying again is something we are good at.

On the other hand, assuming it is coming at all, the economic singularity is coming sooner than the technological singularity. No one knows how long it will take to build an AGI, but it looks tremendously hard. It is probably only a matter of time, but that time may well be quite a few decades. The economic singularity is likely to be with us in two or three decades, and it is likely that large numbers of people will understand that it is coming in around a decade or so, when they see fully autonomous vehicles displacing human drivers. That could lead to a panic, and to some very dangerous populist politics, or worse.

Relatively speaking, then, the technological singularity is more important but less urgent, while the economic singularity is less important but more urgent.

There are many reasons to be excited about artificial intelligence (AI) and what it will do for us and to us. It is already making the world more intelligible, and making our products and services more capable and more efficient.

There are also many reasons to be concerned. People worry about privacy, transparency, security, bias, inequality, isolation, killer robots, oligopoly and algocracy. But none of these concerns refer to phenomena which could, if mishandled, throw our civilisations into reverse gear, or even destroy us completely. The economic and technological singularities could do that if we are foolish and unlucky.

RELINQUISHMENT WON'T WORK

Impressed by the dangers attending the two singularities, it might be a good idea to call a halt to further development of AI research, either permanently, or simply for long enough to allow us to work out how to ensure that both events are beneficial. Unfortunately, this is impossible.

First, we would not know what research to pause. Improvements in the performance of AI comes from many directions: chip design and manufacture, algorithm development, the accumulation and statistical analysis of data, to name but three. Unless we could arrest pretty much all scientific and technological research, we could not be sure that someone, somewhere, was not working on something which will advance AI.

Second, the incentive to develop and deploy a better AI than the competition is literally irresistible. For companies like Google and Facebook, who are leading the way in AI research, it is a matter of critical commercial performance in the short term, and of economic survival in the medium and long term. For military commanders, it is quite literally a matter of life and death. Even if by some miracle all the world's leading politicians could be gathered together and persuaded to sign a joint declaration that all AI research will stop, they would not abide by it. We can all agree that North Korea would cheat, but who would be so naïve as to think that their own government would not do the same?

If it is possible to create an AGI, it will be created – and it will be created as soon as it becomes possible. The same applies to the technologies required to render most humans unemployable.

THE ROLE OF THE TECH GIANTS

Google, Facebook, Amazon, Microsoft, IBM and Apple are shaping the new world we are moving into, along with their Chinese counterparts Baidu, Alibaba and Tencent. Their motivation is partly commercial: they understood sooner than anyone else that AI and related technologies will increasingly provide most of the world's economic value. They are moving aggressively to dominate the AI space, and competing fiercely with each other for talent and market positions.

Although I have no privileged access, it seems to me that many of the leading figures in this industry are also motivated by something else: a belief that the future will be better than today, and an impatience to make it arrive faster.

The idea that AI is improving quickly is now firmly in the public mind. When self-driving cars become common, smartphones are capable of sensible conversations and domestic robots can carry out many of our domestic chores, people will increasingly ask where it is all heading. In the absence of optimistic answers, they will gravitate towards the bad ones, and Hollywood has given us plenty of those.

We need potent new memes, illustrating the current benefits and the future promise of AI. The tech giants are creating this new world; even if only for their own self-preservation, it would be a good idea for them to explain how it is capable of being a glorious new world.

WHAT SHOULD I STUDY?

The question that young people (and their parents) naturally ask about the economic singularity is, how can I best prepare for the economy that we are moving towards? It's an important question for me, too: at the time of writing my son is 16.

The obvious answer is to study computing. Computers are at the heart of the changes sweeping the world in the information revolution, so it has to be valuable to understand how they work and what they can and cannot do. If possible, study machine learning, and in particular, deep learning. At the moment, proficiency with these powerful techniques can earn you a pop star salary, and it seems a safe bet that they will remain important for years to come.

In the long run, however, if the argument of this book is correct, a majority of us are likely to be unemployed. It may well be an advantage for a while to be rich, but if we manage the transition successfully, that may become less important and less worthwhile. And if we don't … well, let's just say we have to.

Beyond the economic singularity, you're going to want to have as rich an interior life as possible, so give yourself as broad an education as you can. Studying the humanities will give you insights into how our minds work. Studying the social sciences will give you insights into how societies work. Studying the hard sciences will give you insights into how the world works. All of these should make what could be a very long life an interesting one.

THE MOST IMPORTANT GENERATIONS

Every generation thinks the challenges it faces are more important than what has gone before. They can't all be correct. American journalist Tom Brokaw bestowed the name 'the greatest generation' on the people who grew up in the Great Depression and went on to fight in World War II. As a late 'baby boomer' myself, I certainly take my hat off to that generation.

Speaking at the United Nations in 1963, John F Kennedy said something which would not sound out of place today: 'Never before has man had such capacity to control his own environment, to end thirst and hunger, to conquer poverty and disease, to banish illiteracy and massive human misery. We have the power to make this the best generation of mankind in the history of the world – or make it the last'.[1]

Today's rising generation is the millennials, born between the early 1980s and the early 2000s. They are also known as generation Y, and the one after them, born from the early 2000s to the early 2020s, is provisionally called generation Z. Let's hope that is not prophetic in a bad way.

The millennials and generation Z have been born at the best time ever to be a human, in terms of life expectancy, health, wealth, access to education information and entertainment. They have also been born at the most interesting time, and the most important. Whether they like it or not, they have the task of navigating us through the economic singularity of mass unemployment, and then, the technological singularity of superintelligence. If they fail, humanity's future could be bleak. But if they succeed, it could be almost incredibly good. They must succeed.

THE FUTURE NEEDS YOU

At the moment, too few of us are paying serious attention to the likely future impact of AI. There is no shortage of stories about it in the media, but they are confused and confusing. Some people appear to be doomsayers (although that is often due to misreporting), and others claim that it is all hype, and there is nothing to get excited about. As a result, most people shudder slightly, shrug their shoulders and get on with the business of living. And who can blame them.

Meanwhile, politicians and policy makers are distracted. The United States is understandably mesmerised by the antics of the 45th president, and in the United Kingdom, Brexit has swallowed the political class whole. Other countries have their own distractions, and the pain of the recession which started in 2008 endures. AI is poised to create the biggest changes humanity has ever been through, and yet it hardly featured at all in recent elections.

This has to change. In particular, we need to start planning for the advent of the economic singularity. The impact of cognitive automation is being felt in modest ways here and there, but the United States, the United Kingdom, and many other leading economies are close to full employment because there is still plenty of work that humans can do. (Some of it doesn't pay very well, but there is work.) This will not last.

Self-driving cars will probably be ready for prime time in five years or so. (It might be seven, or maybe even ten, but Google's Waymo thinks it will be less, and they seem to be in the best position to judge.) At that point, professional drivers will start to be laid off quickly. At the same time, most other sectors of the economy will be seeing the effects. It is imperative that we have a reassuring plan in place by then to let people know that a great future lies ahead.

It could easily take five years to develop that plan and generate a consensus around it. So we have to start now. We need to set up think tanks and research institutes all over the world, properly funded and staffed full-time by smart people with diverse backgrounds and diverse intellectual training. In the context of the importance of the challenge, the resources required are trivial – probably a few tens of millions of dollars – but they are sufficient to require significant political support.

Politicians respond to the public mood. (The most talented ones anticipate it slightly, although they are careful not to get too far ahead of us, or we sack them.) If we demand they pay attention to the coming impact of AI, they will. It is time to make that demand, and you can help. Talk to your friends and colleagues about this: get the conversation going. Insist that your political representatives pay attention.

A wonderful world can be ours if we rise to the challenges posed by the exponential growth of our most powerful technology and navigate the two singularities successfully. Let's grasp that wonderful future!

NOTE

1. http://www.jfklibrary.org/Research/Research-Aids/Ready-Reference/JFK-Quotations.aspx.

Appendix: Other Writers on Technological Unemployment

A.1 PROPHETS OF CHANGE

A.1.1 Martin Ford

Martin Ford is the author of perhaps the best book published so far about the technological unemployment thesis. *The Lights in the Tunnel* (2009) provoked fierce debate, and his follow-up, *Rise of the Robots* (2015) fleshed out his arguments, and responded to the criticism which the first book attracted. Awarding it the 2015 *Financial Times* and McKinsey Business Book of the Year, Lionel Barber, the *Financial Times*' editor, called it 'a tightly written and deeply researched addition to the public policy debate … The judges didn't agree with all of the conclusions, but were unanimous on the verdict and the impact of the book'.

Ford is well placed to talk about what technology will do to the world of work. He has a quarter-century of experience in software design, and he lives and works in Silicon Valley, where he runs a software development company. His writing is calm and measured, with an engaging humility.

Ford opens *Rise of the Robots* with a dramatic illustration of the power of exponential increase – the cumulative doubling which is driving digital innovation. He asks us to consider driving a car at five miles an hour, and then doubling our speed 27 times over. The resulting speed would be 671 million miles an hour – fast enough to travel to Mars in five minutes.[1] This, he points out, is the number of doublings that computer power has gone through since the invention of the integrated circuit in 1958.

The book argues that artificial intelligence (AI) systems are on the verge of wholesale automation of white-collar jobs – jobs involving cognitive skill, such as pattern recognition and the acquisition, processing and transmission of information. In fact, it argues that the process is already under way, and that the US is experiencing a jobless recovery from the Great Recession of 2008 thanks to this automation process. Ford claims that middle-class jobs in the US are being hollowed out, with average incomes going into decline, and inequality increasing. He acknowledges that it is hard to disentangle the impact of automation from that of globalisation and off-shoring, but he remains convinced that AI-led automation is already harming the prospects of the majority of working Americans.

In fact, since Ford's book was published, the US employment figures have improved considerably, and the unemployment rate hovers around 5%, which is considered close to full employment. However, many middle-class Americans do feel squeezed, having been obliged to accept part-time work, or having missed out on wage rises. This suggests that technological unemployment has not yet begun to really bite, but we might be seeing the early warning signs.[2]

Ford pauses to review the prospects for disruption of two sectors of the economy which have so far been relatively unscathed by the digital revolution – education and healthcare. Although there is fierce resistance to the replacement of human activity by AIs in these areas – for instance in essay marking – Ford argues that no industry can ignore for long the benefits of cheaper, faster, more reliable ways of providing their products and services. He goes on to point out that the companies and industries which today are nascent and fast-growing, and tomorrow will be economic giants, are extremely parsimonious employers of humans. AirBnB, the peer-to-peer rooms rental business, for example, achieved a market cap of $20 billion in March 2015 with just 13 employees.

The final chapters of *Rise of the Robots* explore the consequences of the trends which Ford has described. Can an economy thrive and grow if a large minority of people cannot find sufficient work to give themselves and their families a decent life? Would the consequent rise in inequality be economically harmful? More fundamentally, how will these unemployed or underemployed people make ends meet? To Ford, the answer to this last question is clear: governments will need to raise the taxes paid by those who are still working to provide an income for those who are not. But he is acutely aware of the political difficulties that this proposal faces: 'American politicians are terrified to even utter the word "tax" unless it is followed immediately by the word "cut"'.[3]

In fact, Ford seems daunted by the situation: 'The political environment in the United States has become so toxic and divisive that agreement on even the most conventional economic policies seems virtually impossible', he writes. 'A guaranteed income is likely to be disparaged as "socialism"', and 'The decades-long struggle to adopt universal health coverage in the United States probably offers a pretty good preview of the staggering challenge we will face in attempting to bring about any whole-scale economic reform'.

Ford thinks that most people will probably still be able to find some form of paid employment – just not enough to make a decent living. Unwilling to give up on traditional American ideals like the free market, a capitalist economy and indeed the Protestant work ethic, he advocates a universal basic income of only $10,000 a year – a level low enough to leave the incentive to find work in place. Even so, he is pessimistic about the prospect of persuading his fellow Americans to adopt the idea: 'A guaranteed income will probably remain unfeasible for the foreseeable future'.

A.1.2 Andrew McAfee and Frik Brynjolfsson

As a pair of Massachusetts Institute of Technology (MIT) professors,[4] McAfee and Brynjolfsson bring academic credibility to their book on AI automation, *The Second Machine Age*. They have helped to validate the discussion of the possibility of technological unemployment.

Their book (and their argument) is in three parts. The first part (Chapters one through six inclusive) describes the characteristics of what they call the second machine age. They warn readers that their recitation of recent and forthcoming developments may seem like science fiction, and their prose is sometimes slightly breathless: even tenured professors can get excited about the speed of technological change and the wonders it produces.

The second part of the book (Chapters seven through 11) explores the impact of these changes, and in particular two phenomena, which they label 'bounty' and 'spread'. 'Bounty' is the 'increase in volume, variety and quality, and the decrease in cost of the many offerings brought on by technological progress. It's the best economic news in the world today'. This part of the book could have been written by Peter Diamandis, author of *Abundance* and *Bold*, and a leading evangelist for the claim that the exponential growth in computer power is leading us towards utopia.

'Spread' seems to be a synonym for inequality, although the authors are strangely reluctant to use that word.[5] It is 'ever-bigger differences among people in economic success'. This part of the book could have been written by a member of the Occupy movement.[6] 'Spread is a troubling development for many reasons, and one that will accelerate in the second machine age unless we intervene'.

Brynjolfsson and McAfee pose the question whether bounty will overcome the spread. In other words, will we create an economy of radical abundance, in which inequality is relatively unimportant because even though a minority is extraordinarily wealthy, everyone else is comfortably off? Their answer is that current evidence suggests not. Like Martin Ford, they think the American middle class is going backwards financially, and they think this trend will continue unless remedial action is taken.

So, the third and final part of the book discusses the interventions which could maximise the bounty while minimising the spread. In particular, Brynjolfsson and McAfee want to answer a question they are often asked: 'I have children in school. How should I be helping them prepare for the future?'[7] They are optimistic, believing that for many years to come, humans will be better than machines at generating new ideas, thinking outside the box (which they call 'large-frame pattern recognition') and complex forms of communication. They believe that humans' superior capabilities in these areas will enable most of us to keep earning a living, although they think the education system needs to be revamped to emphasise those skills, and downplay what they see as today's overemphasis on rote learning. They praise the Montessori School approach of 'self-directed learning, hands-on engagement with a wide variety of materials … and a largely unstructured school day'. They also have high hopes for digital and distance learning, which use 'digitisation and analytics to offer a host of improvements'.[8]

Brynjolfsson and McAfee offer a series of further recommendations which they say are supported by economists from across the political spectrum: pay teachers more, encourage entrepreneurs, enhance recruitment services, invest in scientific research and infrastructure improvements, encourage immigration by the world's talented migrants, and make the tax system more intelligent.

These seem somewhat unremarkable proposals, although the immigration component could raise hackles in these populist times. The authors acknowledge that their effectiveness may peter out as the 2020s progress, and machines become even smarter. Looking further ahead, they warn against any temptation to try to arrest the progress in AI, and also against any temptation to move away from the tried and tested economic system of capitalism, which they claim (paraphrasing Churchill's quip about democracy again) 'is the worst form of (economy) except for all the others that have been tried'.[9]

The authors are very keen on Voltaire's dictum that 'work saves a man from three great evils: boredom, vice and need'. They are therefore wary of universal basic income, believing that an absence of work will engender boredom and depression. Instead, they argue for a negative income tax, which incentivises work. With a negative income tax of 50%, if you earn a dollar, the government gives you an additional 50 cents. They cast around for ways to keep us all in work, and rather tentatively suggest a range of exotic schemes, such as a cultural movement to prefer goods made by humans rather than machines.

In 2017, Brynjolfsson and McAfee published a new book called *Machine, Platform, Crowd.* Like its predecessor, it is written in an engaging style, and it is an entertaining and illuminating tour of some of the big changes happening in the economy. Although the first third is devoted to AI and related technologies, this time the authors steer even more firmly away from the possibility of technological unemployment. Although they do not say so in the book, it seems they are embarrassed at having raised the issue before. McAfee is quoted as telling an interviewer, 'If I had to do it over again, I would put more emphasis on the way technology leads to structural changes in the economy, and less on jobs, jobs, jobs. The central phenomenon is not net job loss. It's the shift in the kinds of jobs that are available'.[10]

A.1.3 Richard and Daniel Susskind

Father and son team Richard and Daniel Susskind published *The Future of the Professions: How Technology Will Transform the Work of Human Experts* in October 2015. Richard Susskind has impressive credentials, having worked in legal technology since the early 1980s, advised numerous government and industry bodies, and garnered a clutch of honorary fellowships from prestigious universities.[11] Perhaps even more impressive is that he seems to have retained the respect of his subjects while lambasting them as inefficient and doomed to extinction.

The Susskinds describe the 'grand bargain', whereby members of the professions (lawyers, doctors, architects, etc.) enjoy a lucrative monopoly over the provision of certain kinds of advice in return for policing the standard of that provision. They argue that this bargain has broken down, with many professional services now being available only to the rich and well connected. They demonstrate how important this is by illustrating the size of the professions. Healthcare in the US alone now costs $3 trillion a year – more than the GDP of the world's fifth-largest country. The combined revenue of the Big Four accounting firms, at $120 billion, is greater than the GDP of the sixtieth-largest country.

Based on 30 years' experience of the legal industry and backed up by extensive research, the Susskinds paint two scenarios for its midterm future. The first has professionals working closely with technology, their services enhanced as it improves. The second has most or all of the traditional tasks of professionals carried out by machines. The Susskinds believe that this second outcome is the inevitable one, since what the rest of society cares about is not interaction with humans, but getting our legal, medical and other problems sorted out with the minimum of fuss, risk and expense.

The Susskinds keep their focus on the professions, and refrain from making the obvious read-across to the economy as a whole. As a result, they have little to say about universal basic income, or the possibility of society fracturing. But they do note that once machines have taken on responsibility for most or all the tasks previously carried out by human professionals, big questions will be asked about who should own the machines. They don't provide answers to these questions, although they indicate their preference for some form of common ownership which does not involve the state. In this respect, they deserve credit for following the logic of their arguments further than most people writing on the subject.

The book is written with refreshing clarity, precision and felicity of expression – and with such a gloomy message for its target audience, that is probably just as well.

A.1.4 Scott Santens

Scott Santens is a writer and a campaigner for universal basic income, based in New Orleans.[12] He is a moderator of the Reddit basic income page, where he maintains a useful FAQ on the subject.[13] Self-employed since 1997, towards the end of 2015, he managed to procure a basic income for himself based on pledges from others who support his campaign, via the online giving site Patreon.

A.1.5 Jerry Kaplan

Serial entrepreneur Jerry Kaplan co-founded GO Corporation, which was a precursor to smartphones and tablets, and was sold to AT&T. He also co-founded OnSale, an Internet auction site which predated eBay, and was sold for $400 million. He teaches at his alma mater, Stanford University, and writes books, including one called *Humans Need Not Apply*.

Its message is similar to *The Second Machine Age*: AI has reached a tipping point and is becoming powerfully effective. It will disrupt most walks of life, and unless we manage the transition well, the resulting economic instability and growing inequality could be damaging.

Like Ford, Brynjolfsson and McAfee, Kaplan thinks the existing market economy can survive this transition intact.

A.1.6 CGP Grey

Kaplan got the title 'Humans Need Not Apply' from a video of the same name[14] which appeared on the Internet a year before. Posted to YouTube in August 2014 by an Irish-American who goes by the name CGP Grey (his full name is Colin Gregory Palmer Grey[15]), the video attracted over five million views within a year.

The video is well-produced, engaging and persuasive. It contains plenty of technological eye candy, and makes its points in punchy sound-bites – ideal for today's short attention spans. Unlike the books described previously, it offers no solutions to the problems raised by AI and robotic automation, but – also unlike them – it suggests that capitalism cannot cope with what is coming.

A.1.7 Gary Marcus

A psychology professor at New York University, Gary Marcus has taken an intense interest in AI, and where it is leading us. In February 2015, he told a CBS interviewer: 'Eventually I think most jobs will be replaced, like 75%–80% of people are probably not going to work for a living ... There are a few people starting to talk about it'.[16]

A.1.8 Federico Pistono

Federico Pistono is a young Italian lecturer and social entrepreneur. He attracted considerable attention with his 2012 book *Robots Will Steal Your Job, But That's OK*. A range of eminent people, including Google's Larry Page, were drawn to its optimistic and discursive style.

After making a forceful case that future automation will render most people unemployed, Pistono argues that there is no need to worry. Much of the book is taken up with musing on the nature of happiness – the word features in the titles of a quarter of its chapters. He is hopeful that we will all discover that the pursuit of happiness through material goods is a fool's errand, and he argues that salvation lies in downsizing. He offers the example of his own family, living in Northern Italy. They spend $45,000 a year, but by getting rid of two of their three cars, growing their own food, and generating their own electricity, they can reduce this to $29,000 a year.

He also urges us all to educate ourselves – and encourage everyone else to do likewise – but more for personal fulfilment than in a vain attempt to remain employable.

A.1.9 Andy Haldane

As the chief economist of the Bank of England, Andy Haldane isn't the most obvious person to be found musing about the benefits of universal basic income. But that is exactly what he did in a speech at the Trades Union Congress in November 2015.[17] He wondered whether the displacement effect of automation, whereby jobs are destroyed, might start to outweigh the compensation effect, whereby automation raises productivity sufficiently to generate more demand and thus work.

In his speech, Haldane avoided giving a definitive answer to the question of whether we are nearing 'peak human', but he raised many of the concerns explored in this book. He presented an estimate prepared by the Bank of England of the likelihood of automation of the jobs in a range of economic sectors in the UK, adapted from the estimates produced for the United States by Frey and Osborne of the Oxford Martin School (of which, more will follow). The Bank estimated the UK's situation as slightly less alarming than that of the United States, but not much. It found that roughly a third of jobs have a low probability of being automated out of existence, another third have a medium probability and the final third have a high probability. Haldane avoided putting a specific timescale on this, and also avoided saying what would happen after that undisclosed period.

A.1.10 Martin Wolf

As the main financial columnist and associate editor at the *Financial Times*, Martin Wolf is the very epitome of a City establishment figure. He was described by US Treasury Secretary Larry Summers as 'probably the most deeply thoughtful and professionally informed economic journalist in the world'.[18] Although the credit crunch and subsequent recession have rekindled his youthful enthusiasm for Keynesian economics, it is still a surprise to read him advocating income redistribution and universal basic income, as he did in this article from February 2014:

'If Mr Frey and Prof Osborne [see subsequent sections] are right [about automation]... we will need to redistribute income and wealth. Such redistribution could take the form of a basic income for every adult, together with funding of education and training at any stage in a person's life. The revenue could come from taxes on bads (pollution, for example) or on rents (including land and, above all, intellectual property). Property rights are a social creation. The idea that a small minority should overwhelming(ly) benefit from new technologies should be reconsidered. It would be possible, for example, for the state to obtain an automatic share in the income from the intellectual property it protects'.[19]

A.2 ACADEMICS, CONSULTANTS AND THINK TANKS

Numerous reports have been written about technological unemployment by academic organisations, consultancies and think tanks. I have described some of the better-known ones here. Sometimes they reserve judgement or sit on the fence, but as far as possible, I present them in order of increasing scepticism about the proposition of widespread unemployability.

A.2.1 Frey and Osborne

Carl Benedikt Frey and Michael Osborne are the directors of the Oxford Martin Programme on Technology and Employment.[20] Their 2013 report *The Future of Employment: How Susceptible Are Jobs to Computerisation* has been widely quoted. Its approach to analysing US job data has since been used by others to analyse job data from Europe and Japan.

The report analyses 2010 US Department of Labor data for 702 jobs, and in a curious blend of precision and vagueness, concludes that '47% of total US employment is in the high-risk category, meaning that associated occupations are potentially automatable over some unspecified number of years, perhaps a decade or two'. Nineteen per cent of the jobs were found to be at medium risk and 33% at low risk. Studies which have extended these findings to other territories have yielded broadly similar results.

The methodology overlays rigour on guesswork. Seventy of the jobs were categorised in a brainstorming session, and these categorisations were then extended to the other 632 jobs using calculations which will mystify anyone with only school-level maths, including a statistical tool called Gaussian process classifiers. But it would be unfair to criticise the report for lack of rigour. Forecasting is not an exact science; the authors adopted the most scientific approach they could devise, and made no attempt to hide its subjective elements.

As well as sounding the alarm about the possibility of technological unemployment, the report suggests that the 'hollowing out' of middle-class jobs will stop. A 2003 paper by David Autor (of whom, more follows) observed that income has increased for high earners and (albeit less rapidly) for low earners, but stagnated for medium-level earners. Maarten Goos and Alan Manning characterised this hollowing out as the favouring of 'Lovely and Lousy jobs'. Frey and Osborne argue that in the future, susceptibility to automation will correlate negatively with income and educational attainment, so the Lousy jobs will also disappear. They suggest that people will have to acquire creative and social skills to stay in work, but they don't appear to think that many of us will be able to change the fate that our employment history has assigned to us.

Frey and Osborne followed the 2013 report with another in February 2015, written in collaboration with senior bankers from Citibank. It provides insights into the impact of automation in a number of industry sectors, including stock markets, where the move from trading floors to digital exchanges reduced headcount by 50%. At first glance, it is surprising to see bankers suggesting increased taxation to provide income for the unemployed, but it also seems they have little faith in it happening: 'Such changes in taxation would seem sensible to us, but they would also be a reversal of the trends of the last few decades'. They don't hold out much more hope for their other principal suggested remedy: 'Education alone is unlikely to solve the problem of surging inequality, [but] it remains the most important factor'.

A.2.2 Gartner

Gartner is the world's leading technology market research and advisory consultancy. At its annual conference in October 2014, its research director Peter Sondergaard declared that one in three human jobs would be automated by 2025.[21] 'New digital businesses require less labor; machines will make sense of data faster than humans can'. He described smart machines as an example of a 'super class' of technologies which carry out a wide variety of tasks, both physical and intellectual. He illustrated the case by pointing out that machines have been grading multiple-choice examinations for years, but they are now moving on to essays and unstructured text.

A.2.3 The Millennium Project

The Millennium Project was established in 1996 by a coalition of UN organisations and US academic research bodies. Its *2015–16 State of the Future* contained a section on the future of work based on a poll of 300 experts from around the world. Although they mostly thought that technology would impact employment significantly, their collective estimates for long-term unemployment were relatively conservative. They expected global unemployment to reach only 16% in 2030, and just 24% in 2050.

A.2.4 Pew Research Center

The Pew Research Center published a report entitled *AI, Robotics, and the Future of Jobs*[22] in November 2014. The Center is part of the Pew Charitable Trusts, established in 1948 with over $5 billion bequeathed by descendants of the founder of Sun Oil; the Center is the third-largest think tank in the United States.

The Center sent a questionnaire to 12,000 selected experts and interested members of the public (mostly but not entirely American), and received 1,900 responses to the question 'Will networked, automated, artificial intelligence (AI) applications and robotic devices have displaced more jobs than they have created by 2025?'. A slight majority (52%) said no, arguing that technology has always created more jobs than it has destroyed, that it is not advancing fast enough to destroy so many jobs, and that regulatory intervention would stop it if necessary.

The 48% who thought there would be a net loss of jobs believed that the process was already in train, but that it would get much worse, and that inequality would become a severe problem as a result.

Both sides thought that the education system is doing a poor job of preparing young people for the new world of work, and also that the future of employment is not preordained, but is susceptible to good policy.

A.2.5 Fundacion Innovacion BankInter

BankInter, based in Madrid, is one of the largest banks in Spain. In 2003, it established a foundation to promote the creation of sustainable wealth in Spain through innovation and entrepreneurship. One of the foundation's main activities is organising the Future Trends Forum, an international think tank which periodically gathers together a group of experts to discuss an important topic, and then produces reports and videos based on the conclusions of those discussions.

In June 2015, I took part in a meeting of the Future Trends Forum entitled 'The Machine Revolution', which addressed 'how technological developments (internet, robotics, artificial intelligence, etc.) will boost employment and labour markets in the next decade'. Compered by Chris Meyer, author of *Standing on the Sun*, the delegates were a mixture of senior government officials from around the world, academic and commercial economists, investors and writers.

When the 34 experts at the meeting were asked whether we thought structural unemployment was a likely result, a slight majority said not. Towards the end of the meeting, we each contributed two predictions to a collective timeline, which appears at the end of the report.[23]

A.2.6 McKinsey

The world's most prestigious management consultancy firm weighed in on the subject of technological unemployment with an article published in its quarterly magazine in November 2015 entitled 'Four Fundamentals of Workplace Automation'.[24] Billed as an interim report of an ongoing research project, its central argument was that instead of asking which jobs can be and will be automated, we should ask which *tasks* will be automated. Few people, it claimed, will find that their entire job disappears, but as much as 45% of the tasks people do at work can be automated with technology that is currently available.

The McKinsey consultants identified 2,000 different 'activities' (e.g. greeting customers, demonstrating product features) for a selection of 'occupations' (e.g. retail salesperson), and assessed which activities required the 18 'capabilities' which they deem susceptible to automation (e.g. understanding natural language, generating natural language, retrieving information).

They noted that the level of automatability will rise as machines become more capable. For instance, if and when machines equal the median human level of natural language comprehension, then the proportion of tasks which can be automated will rise from 45 to 58%.

At the time of publication, the authors concluded that only 5% of jobs were capable of being fully automated, but 60% of jobs could have 30% of their constituent activities automated. But rather than leading to a 30% headcount reduction, with the other 70% of activities being smeared among the remaining employees, they expect people to become more productive, as machines augment human performance.

Very highly paid jobs tended to have fewer automatable activities (e.g. 20% for CEOs), but among low- and medium-paid jobs there was a fairly even spread. The consultants found that only 4% of the work activities carried out in the United States require creativity at the median human level, and only 29% require a median human level of emotion sensing. Optimistically, they concluded from this that automation will enable humans to do better and more interesting work. Interior designers, for instance, 'could spend less time taking measurements, developing illustrations and ordering materials, and more time developing innovative design concepts'.

Finally, McKinsey suggested that senior managers should increasingly pay close attention to the type, direction and potential of automation within their industry, as it will become a more and more important source of competitive advantage.

A.2.7 A Swelling Chorus

The reports described previously are a selection of the most prominent ones published so far on the subject of technological unemployment. They are not the only ones, and more are being produced every month – sometimes every week. There is no clear consensus about the likely impact on joblessness of machine intelligence in the coming years and decades. Nevertheless, the theme has an increasingly high profile in the media – it was a focus of the annual gathering of the super-rich and powerful in the ski resort of Davos in January 2016.

In the next section we will see that some people are firmly convinced it is a myth.

A.3 SCEPTICS

In this section we meet a selection of writers who are sceptical about the prospect of technological unemployment. They argue that it is all just a revival of the Luddite fallacy.

A.3.1 David Autor

David Autor is a professor of economics at MIT. As noted previously, he sounded the alarm in a 2003 paper about the 'hollowing out' of middle-class jobs in the United States – the fact that income has increased for high earners and (albeit less rapidly) for low earners, but stagnated for medium-level earners.

In an interview in October 2015,[25] he gave three reasons why he thinks that some observers have been unduly pessimistic, even hysterical, about the likelihood of job destruction. One is that machines complement and augment humans: they always have, and there is no reason to think that is about to change. The second is that machines increase productivity, which creates wealth, consumption; and demand, which creates more jobs.

The third reason is that humans are creative and ingenious. There are many important businesses and activities now that could not have been imagined 50 years ago. In fact, Autor accuses Martin Ford of arrogance in writing off human ingenuity.

In an article for the *Journal of Economic Perspectives* (summer 2015) entitled 'Why Are There Still So Many Jobs?',[26] Autor forecasts that people will retain a comparative advantage in so-called 'human' attributes, such as interpersonal interaction, flexibility, adaptivity. He argues that many jobs – like radiologist – combine these with the routine, predictable tasks where computers win. Autor believes it will not be possible to separate these two types of tasks, so humans will continue to carry out the whole bundle.

More generally, Autor is also one of those who believe the rate of change today is over-hyped. He believes that the effect of Moore's law is substantially muted by regulatory and social frictions which slow down the adoption of new technologies, and he also argues that many technological advances simply don't translate into tangible improvements in the real world. For instance, he accepts that his current computer is a thousand times faster than the one he used a few years ago, but he suspects it only makes him 20% more productive. It may be true, he teases, that a new washing machine has more processing power than NASA used to send Neil Armstrong to the moon in 1969, but the washing machine is still not going to the moon.

He derides some of the heralded achievements of AI researchers, arguing for instance that self-driving cars do not emulate human drivers, but instead rely on precise maps of the terrain which have to be prepared before the journey starts. This makes them less flexible than humans, and not fit to be released into the wild without human escorts.

Although Autor is broadly optimistic about our future, he believes that much depends on the decisions that we take. 'If machines were in fact to make human labour superfluous, we would have vast aggregate wealth but a serious challenge in determining who owns it and how to share it'. He points out that Norway and Saudi Arabia both enjoy economic abundance (thanks to oil rather than AI), but they use it very differently. Norwegians, he says, work few hours per day and are generally happy; Saudis import 90% of their labour and nurture terror.

A.3.2 Robin Hanson

Robin Hanson is an associate professor of economics at George Mason University, in Virginia, United States. Like David Autor, Hanson castigates Martin Ford for inappropriate motives, but whereas Autor accuses

Ford of arrogance, Hanson alleges dishonesty: 'In the end, it seems that Martin Ford's main issue really is that he dislikes the increase in inequality and wants more taxes to fund a basic income guarantee. All that stuff about robots is a distraction'.[27]

After a few more jibes, Hanson addresses Ford's actual thesis. He starts by admitting that 'Ford is correct that, in the long run, robots will eventually get good enough to take pretty much all jobs. But why should we think something like that is about to happen, big and fast, *now*?' He attributes four arguments to Ford, and makes short work of the first three. The first is the Frey and Osborne study we reviewed in Section A.2.1, which Hanson dismisses as subjective. The second argument is the decline in labour's share of income since 2000, which Hanson replies could be caused by numerous other factors rather than technological automation. The third argument is the rapid fall in computer prices, which Hanson says has yet to cause any detectable unemployment.

'And then there is Ford's fourth reason: all the impressive computing demos he has seen lately'. Hanson is referring, of course, to Google's self driving cars, real-time machine translation systems, DeepMind's Atari-playing system and so on. Hanson is less impressed by these demonstrations of rapidly improving AI: 'We do expect automation to take most jobs eventually, so we should work to better track the situation. But for now, Ford's reading of the omens seems to me little better than fortune telling with entrails or tarot cards'.

Having unburdened himself of this cynicism, Hanson proceeds to offer a constructive suggestion. He advocates forecasting by means of prediction markets, where people place bets on particular economic or policy outcomes, like the level of unemployment at some future date. He argues that prediction markets give us a financial stake in being accurate when we make forecasts, rather than just trying to look good to our peers.

A.3.3 Tyler Cowen

A professor at George Mason University and co-author of an extremely popular blog, Tyler Cowen was New Jersey's youngest-ever chess champion. He is a man with prodigiously broad knowledge and interests, and although he proposes some key ideas forcefully, there is always some nuance, and he dislikes simplistic and modish solutions. In two recent books, *The Great Stagnation* (2011) and *Average is Over* (2014), he paints a picture of America's future which is slightly depressing, but not apocalyptic. He is alive to the prospect of dramatically improved AI, and the effect it will have on employment. But he does not think widespread permanent unemployment will be one of its results.

For several years, Cowen has championed the claim that the US economy is hollowing out. He expects automation to continue this trend, perhaps to accelerate it. In an article for *Politico* magazine,[28] he wrote:

'I imagine a world in which, say, 10 to 15 percent of the citizenry ... is extremely wealthy and has fantastically comfortable and stimulating lives, equivalent to those of current-day millionaires, albeit with better health care. Much of the rest of the country will have stagnant or maybe even falling wages in dollar terms'. This grim outlook for the majority is softened because 'they will have a lot more opportunities for cheap fun and cheap education (thanks to) all the free or nearly free services that modern technology makes available'. But there is a sting in the tail for the real underclass. They, he says, 'will fall by the wayside'.

Cowen does not expect a universal basic income to be required. Nor does he expect riots. One reason is that the US population will be older: 'By 2030, about 19 percent of the US population will be over 65; in other words, we'll be as old as Floridians are today'. Floridians are a conservative bunch, not given to mayhem. Another is that people will increasingly cluster geographically according to income. Few people in the poorer 85% will live in the hothouse cities of San Francisco and New York, and they will not have the wealth of Manhattan waved in their faces. And perhaps most important, the masses will inure themselves with the opiates of free entertainment and social media.

A.3.4 Geoff Colvin

Geoff Colvin is an editor at *Fortune* magazine and one of America's most experienced and respected journalists. In August 2015, he published *Humans Are Underrated: What High Achievers Know That Brilliant Machines Never Will*. His previous book, *Talent is Over-Rated*' (2006) advanced the proposition that hours of dedicated practice trumps talent in most endeavours, and it was an international best-seller.

His new book accepts that for the first time, technology may be reducing total employment rather than increasing it, but is sceptical for two reasons in particular. First, Colvin argues that because it is so hard to foresee the new types of jobs that are created when economies shift (just as web development and social media marketing were hard to foresee), we underestimate how many of them there will be.

Colvin believes that skills of deep human interaction – empathy, storytelling and the ability to build relationships – will become far more valuable in the future, and many people will be able to prosper by bringing those skills into the evolving economy. 'We're hard-wired by 100,000 years of evolution to value deep interaction with other humans (and not with computers). Those wants won't be changing anytime soon'.[29]

NOTES

1. This depends on the two planets being pretty much as close as they ever get.
2. http://fortune.com/2015/11/10/us-unemployment-rate-economy/.
3. This and the other quotes in this paragraph and the next one are from Chapter 10: Towards a New Economic Paradigm.
4. Brynjolfsson is the director of the MIT Center for Digital Business and McAfee is a principal research scientist there.
5. The word 'inequality' crops up 42 times in the book, including in the titles of sources, but the authors never explicitly connect it with 'spread'.
6. The loosely organised protest organisation that sprang up after the 2008 credit crunch to campaign against inequality.
7. Chapter 12: Learning to Race with the Machines: Recommendations for Individuals.
8. Chapter 13: Policy Recommendations.
9. Chapter 14: Long-Term Recommendations.
10. https://www.wired.com/2017/08/robots-will-not-take-your-job.
11. http://www.susskind.com/.
12. http://www.scottsantens.com/.
13. https://www.reddit.com/r/BasicIncome/ and https://www.reddit.com/r/basicincome/wiki/index.
14. https://www.youtube.com/watch?v=7Pq-S557XQU.
15. https://www.youtube.com/watch?v=C5MVXdg6nho.
16. http://www.cbsnews.com/videos/how-technology-may-change-our-labor-and-leisure/.
17. http://www.bankofengland.co.uk/publications/Pages/speeches/2015/864.aspx.
18. https://newrepublic.com/article/69326/call-the-wolf.
19. http://www.ft.com/cms/s/0/dfe218d6-9038-11e3-a776-00144feab7de.html#axzz3stkJb1V2.
20. The Programme was established in January 2015 with funding from Citibank, one of the largest financial institutions in the world. The Oxford Martin School was set up as part of Oxford University in 2005, as an institution dedicated to understanding the threats and opportunities facing humanity in the twenty-first century. It is named after James Martin, a writer, consultant and entrepreneur, who founded the school with the largest donation ever made to the university – which was no mean feat given that Oxford was founded 1,000 years ago, and is the oldest university in the world (after Bologna in Italy).

21. http://www.computerworld.com/article/2691607/one-in-three-jobs-will-be-taken-by-software-or-robots-by-2025.html.

22. http://www.pewinternet.org/2014/08/06/about-this-report-and-survey-2/?beta=true&utm_expid=53098246-2.Lly4CFSVQG2lphsg-KopIg.1&utm_referrer=https%3A%2F%2Fwww.google.co.uk%2F.

23. https://www.fundacionbankinter.org/web/fundacion-bankinter/ficha-documento?param_id=173404#_48_INSTANCE_av33_%3Dhttps%253A%252F%252Fwww.fundacionbankinter.org%252Fweb%252Fglobal-site%252F-%252Fthe-machine-revolution%253F.

24. http://www.mckinsey.com/insights/business_technology/four_fundamentals_of_workplace_automation.

25. http://www.socialeurope.eu/2015/10/the-limits-of-the-digital-revolution-why-our-washing-machines-wont-go-to-the-moon/

26. https://www.aeaweb.org/articles.php?doi=10.1257/jep.29.3.3

27. https://reason.com/archives/2015/03/03/how-to-survive-a-robot-uprisin.

28. http://www.politico.com/magazine/story/2013/11/the-robots-are-here-098995.

29. http://www.forbes.com/sites/danschawbel/2015/08/04/geoff-colvin-why-humans-will-triumph-over-machines/2/.

Index